同多化合物概论

牛景杨　王敬平　编著

化学工业出版社
·北京·

内容提要

同多化合物是一类由前过渡金属构成的具有明确结构的多核配合物,在催化、材料及药物化学领域具有广阔的应用前景。

本书共分 11 章。第 1 章对同多化合物作了大致的介绍;第 2 章到第 10 章,分别介绍了从二金属氧簇到十金属氧簇的合成、结构、性质及相关化合物的研究;第 11 章集中介绍了金属氧簇中金属原子个数超过 10 个的高核簇的合成及结构。

本书可供从事多金属氧簇及相关领域的研究人员参考,也可作为相关专业研究生用书。

图书在版编目(CIP)数据

同多化合物概论/牛景杨,王敬平编著.—北京:化学工业出版社,2019.11
ISBN 978-7-122-35364-1

Ⅰ.①同… Ⅱ.①牛… ②王… Ⅲ.①过渡金属化合物-概论 Ⅳ.①O614

中国版本图书馆 CIP 数据核字(2019)第 215651 号

责任编辑:成荣霞　　　　　　　　　　　文字编辑:李　玥
责任校对:刘　颖　　　　　　　　　　　装帧设计:王晓宇

出版发行:化学工业出版社 (北京市东城区青年湖南街 13 号　邮政编码 100011)
印　　装:北京虎彩文化传播有限公司
710mm×1000mm　1/16　印张 14½　字数 268 千字　2020 年 11 月北京第 1 版第 1 次印刷

购书咨询:010-64518888　　　售后服务:010-64518899
网　　址:http://www.cip.com.cn
凡购买本书,如有缺损质量问题,本社销售中心负责调换。

定　价:128.00 元　　　　　　　　　　　　　　　版权所有　违者必究

前言
Preface

多金属氧簇（多酸）是一大类由前过渡金属以氧桥联构成的多核化合物，由于其在催化、材料及药物等领域具有重要应用价值和潜在应用前景而受到了广泛的关注，成为无机化学非常活跃的研究领域，也促进了催化、材料、药物及环境等相关学科的研究。根据多金属氧簇分子内组成元素的不同，该类化合物可分为同多化合物和杂多化合物，其核心的差别在于同多化合物是由同种含氧酸根缩合而形成同多酸根阴离子，进而构成同多化合物；杂多化合物是由不同的含氧酸根缩合而形成杂多阴离子，进而构成杂多化合物。尽管二者密切相关而且统称为多金属氧簇或多酸，但二者在合成方法、组成、结构及性质上的差别还是客观存在的。笔者曾结合自己的科研及研究生教学经验，写过一本系统介绍杂多化合物的《杂多化合物概论》，此后就一直想就同多化合物也写一本类似的书来系统介绍同多化合物，但由于同多化合物结构类型繁杂、研究报道零散，很难决定取舍以形成清晰的脉络，虽然着手较早，但一直拖延至今。

本书是在笔者多年从事同多化合物合成与性质研究及讲授研究生课程的基础上，结合当前同多化合物研究的最新进展编写而成的。

在编写的过程中得到了本课题组的张东娣、马鹏涛、张超几位老师的大力支持和帮助，研究生梁志结、万蓉等同学协助查阅了文献并帮助录入了书稿，胡枫同学帮助制作了部分插图，在此一并表示衷心感谢。

笔者力图全面地描述同多化合物的发展进程和最新研究成果，但本领域研究时间跨度很大，研究进展日新月异，文献语言种类繁多，尽管笔者对文献进行过多次检索，但限于水平，漏检及取舍不当在所难免，欢迎读者批评指正。

编者

目录
Contents

第1章　绪论 / 001

第2章　二金属氧簇 / 003

2.1　二钒氧簇 / 003
　2.1.1　合成 / 003
　2.1.2　结构 / 004
　2.1.3　相关化合物 / 004
2.2　二钼氧簇 / 005
　2.2.1　合成 / 005
　2.2.2　谱学表征 / 005
　2.2.3　结构 / 006
　2.2.4　相关化合物 / 006
参考文献 / 009

第3章　三金属氧簇 / 013

3.1　三钒氧簇 / 013
　3.1.1　合成 / 013
　3.1.2　谱学表征 / 014
　3.1.3　结构 / 015
3.2　三钼钨氧簇 / 016
参考文献 / 016

第4章　四金属氧簇 / 020

4.1　四钒氧簇 / 020
　4.1.1　合成 / 020
　4.1.2　谱学表征 / 021
　4.1.3　结构 / 021
　4.1.4　相关化合物 / 022
4.2　四钼氧簇 / 024
4.3　四钨氧簇 / 028
参考文献 / 031

第 5 章　五金属氧簇 / 037

5.1　五钒氧簇 / 037
- 5.1.1　合成 / 037
- 5.1.2　谱学表征 / 038
- 5.1.3　晶体结构 / 038

5.2　五钼氧簇 / 038
- 5.2.1　合成 / 038
- 5.2.2　结构 / 039
- 5.2.3　相关化合物 / 039

5.3　五钨氧簇 / 043
- 5.3.1　合成 / 044
- 5.3.2　结构 / 044
- 5.3.3　相关化合物 / 044

参考文献 / 048

第 6 章　六金属氧簇 / 051

6.1　六钒氧簇 / 051
- 6.1.1　合成 / 051
- 6.1.2　结构 / 052

6.2　六铌氧簇 / 058
- 6.2.1　合成 / 059
- 6.2.2　谱学表征 / 059
- 6.2.3　结构 / 060
- 6.2.4　相关化合物 / 061

6.3　六钽氧簇 / 064
- 6.3.1　合成 / 064
- 6.3.2　相关化合物 / 065

6.4　六钼氧簇 / 065
- 6.4.1　合成 / 065
- 6.4.2　结构 / 066
- 6.4.3　振动光谱 / 068
- 6.4.4　衍生物 / 071

6.5　六钨氧簇 / 077
- 6.5.1　合成 / 078
- 6.5.2　谱学表征 / 078
- 6.5.3　结构研究 / 078

  6.5.4 衍生物 / 079
 参考文献 / 082

第 7 章 七金属氧簇 / 094

 7.1 七铌氧簇 / 094
 7.2 七钼氧簇 / 095
  7.2.1 合成 / 095
  7.2.2 谱学表征 / 095
  7.2.3 晶体结构 / 096
  7.2.4 性质研究 / 097
  7.2.5 相关化合物 / 099
 7.3 七钨氧簇 / 100
  7.3.1 合成 / 100
  7.3.2 结构 / 100
  7.3.3 相关化合物 / 100
 参考文献 / 101

第 8 章 八金属氧簇 / 106

 8.1 八金属氧簇 M_8O_{26} / 106
  8.1.1 α 异构体 / 106
  8.1.2 β 异构体 / 111
  8.1.3 γ 异构体 / 125
  8.1.4 δ 异构体 / 133
  8.1.5 ε 异构体 / 135
  8.1.6 ζ 异构体 / 137
  8.1.7 η 构型 / 138
  8.1.8 θ 构型 / 138
  8.1.9 八钼氧簇异构体比较 / 139
 8.2 八金属氧簇 M_8O_{30} / 139
  8.2.1 八钼氧簇 / 139
  8.2.2 八钨氧簇 / 140
 参考文献 / 141

第 9 章 九金属氧簇 / 149

 参考文献 / 150

第 10 章 十金属氧簇 / 151

 10.1 十钒氧簇 / 151

  10.1.1 合成 / 151
  10.1.2 谱学表征 / 152
  10.1.3 晶体结构 / 152
  10.1.4 相关化合物 / 154
 10.2 十铌氧簇 / 165
  10.2.1 合成 / 165
  10.2.2 谱学表征 / 165
  10.2.3 晶体结构 / 165
  10.2.4 相关化合物 / 166
 10.3 十钼氧簇 / 167
  10.3.1 合成 / 167
  10.3.2 晶体结构 / 167
  10.3.3 相关化合物 / 168
 10.4 十钨氧簇 / 170
  10.4.1 合成 / 170
  10.4.2 谱学表征 / 171
  10.4.3 晶体结构 / 172
  10.4.4 相关化合物 / 172
参考文献 / 176

第11章 高聚金属氧簇 / 183

 11.1 钒氧簇 / 183
  11.1.1 十二钒氧簇 / 183
  11.1.2 十三钒氧簇 / 184
  11.1.3 十四钒氧簇 / 185
  11.1.4 十五钒氧簇 / 186
  11.1.5 十六钒氧簇 / 189
  11.1.6 十七钒氧簇 / 191
  11.1.7 十八钒氧簇 / 192
  11.1.8 十九钒氧簇 / 192
  11.1.9 二十二钒氧簇 / 194
  11.1.10 三十四钒氧簇 / 195
 11.2 铌氧簇 / 195
  11.2.1 二十铌氧簇 / 195
  11.2.2 二十四铌氧簇 / 195
  11.2.3 二十七铌氧簇和三十一铌氧簇 / 196
 11.3 钼氧簇 / 197

11.3.1　十三钼氧簇 / 197
　　11.3.2　十六钼氧簇 / 198
　　11.3.3　三十六钼氧簇 / 199
　　11.3.4　三十七钼氧簇 / 200
　　11.3.5　四十二钼氧簇 / 201
　　11.3.6　四十六钼氧簇 / 201
　　11.3.7　五十四钼氧簇 / 201
　　11.3.8　其他高核钼氧簇 / 202
　11.4　钨氧簇 / 205
　　11.4.1　十一钨氧簇 / 205
　　11.4.2　十二钨氧簇 / 205
　　11.4.3　十九钨氧簇 / 207
　　11.4.4　二十二钨氧簇 / 207
　　11.4.5　二十四钨氧簇 / 207
　　11.4.6　二十八钨氧簇 / 208
　　11.4.7　三十四钨氧簇 / 208
　　11.4.8　三十六钨氧簇 / 209
　参考文献 / 209

索引 / 220

第 1 章

绪 论

同多化合物是由同种含氧酸根缩合而成的同多阴离子构成的化合物,与杂多化合物一样,均为多金属氧簇的重要组成。

首例同多化合物的研究报道已很难考证,但现有文献表明,由于自然界存在同多阴离子——十钒氧簇阴离子构成的矿物:橙钒钙矿($Ca_3V_{10}O_{28} \cdot 17H_2O$)、水钒镁钾石($K_2Mg_2V_{10}O_{28} \cdot 17H_2O$)和水钒镁钠石($Na_4MgV_{10}O_{28} \cdot 24H_2O$),因此,人们在很早就开始了对十钒氧簇阴离子的研究,但当时的研究仅限于对其性质的探索,而对其结构却很少涉及。1903 年,Dullberg 等人提出在溶液中存在 $HV_6O_{17}^{3-}$ 六钒氧簇阴离子,但直到现在也没有得到具有独立结构的六钒氧簇阴离子形成的化合物。1952 年,Linqvist 通过对 $Na_7HNb_6O_{19} \cdot 16H_2O$ 的研究,首次描述了六铌氧簇阴离子的结构,这是首例报道的结构明确的同多阴离子构成的化合物。随着研究工作的不断深入,不同核数、不同结构类型的新型同多金属氧簇阴离子构成的化合物不断被报道出来。迄今,已报道的同多金属氧簇按所含金属核数划分有二金属氧簇、三金属氧簇、四金属氧簇、五金属氧簇、六金属氧簇、七金属氧簇、八金属氧簇、九金属氧簇、十金属氧簇以及更高金属核数的同多金属氧簇阴离子。到目前为止,已报道的含有最多金属核数的同多金属氧簇化合物为 $Na_{48}[H_xMo_{368}O_{1032}(H_2O)_{240}(SO_4)_{48}]Ca \cdot$

$1000H_2O$，其阴离子包含了368个钼原子。

在相同金属核数的同多金属氧簇中，氧原子个数的不同将会导致金属氧簇结构的不同及性质的改变；在原子组成相同的金属氧簇中，还可能存在同分异构体；而且还存在大量的配位聚合物，这就构成了目前同多化合物领域丰富多彩的结构化学。

由于同多金属氧簇结构的多样性，本书在编写过程中材料取舍非常踌躇，后经反复斟酌，形成如下编写思路。

（1）分类

由于同多阴离子的结构类型很丰富，很多阴离子结构难以归属于某一种结构类型，因此，在本书中，以同多阴离子中金属原子的个数为分类依据，具有相同金属原子个数的阴离子归为一类，其组成和结构的差异作为同分异构体或相关化合物介绍。

（2）范围

由于本书以同多阴离子内金属原子个数作为分类依据，因此，本书中只介绍了阴离子内金属原子个数明确的化合物，而金属原子个数难以确定的化合物，例如多维结构化合物不在本书介绍之列。另外，本书介绍的同多化合物主要是由过渡元素钒、铌、钼和钨形成的，而其他元素构成的少量化合物不在介绍范围内。

（3）历史

在编写过程中，我们曾试图考证清楚每一种结构类型的同多阴离子首次报道的时间以及形成的过程，但后来发现，由于该类化合物的历史悠久以及文献有限，这一目标难以实现，因此在书中只对部分可以查到的作了简介。

第2章 二金属氧簇

二聚金属氧簇是结构最简单的多金属氧簇。有关钒、铌、钽、钼和钨二聚金属氧簇的理论探讨有多篇报道[1-4],但到目前为止,能形成稳定的独立结构并从溶液中分离出来的只有二钒氧簇和二钼氧簇形成的化合物[5-9]。尽管人们在合成二钨氧簇方面也做了多次努力[10,11],但至今仍未得到具有独立结构的二钨氧簇的稳定化合物[1,12]。而对铌、钽二聚金属离子的报道仅限于理论探讨[4]。下面就对二聚金属氧簇及其相关化合物作一介绍。

2.1 二钒氧簇

2.1.1 合成

早在1933年Jander用pH滴定及溶液中离子扩散系数测定法研究多钒酸盐溶液体系时就提出在VO_4^{3-}的酸性溶液中存在聚合度为$V_2O_7^{4-}$的多钒酸根离子[13]。1945年Souchay用与上述类似的实验方法,并结合溶解度法及溶液依数性,更精确地研究了多钒酸盐溶液体系,认为在钒酸盐的酸性溶液中存在如下平衡关系[14]:

$$[VO_4]^{3-} + H^+ \rightleftharpoons HVO_4^{2-}$$
$$2HVO_4^{2-} \rightleftharpoons [V_2O_7]^{4-} + H_2O$$
$$3HVO_4^{2-} + 3H^+ \rightleftharpoons [V_3O_9]^{3-} + 3H_2O$$
$$2[V_3O_9]^{3-} + 2H^+ \rightleftharpoons [V_6O_{17}]^{4-} + H_2O$$
$$[V_6O_{17}]^{4-} + H^+ \rightleftharpoons H[V_6O_{17}]^{3-}$$
$$H[V_6O_{17}]^{3-} + H^+ \rightleftharpoons H_2[V_6O_{17}]^{2-}$$

后来的研究结果支持了在钒酸盐的酸性溶液中存在二聚钒氧簇离子 $V_2O_7^{4-}$ 的结论[15-18]。尽管有很多化学计量比为 $V_2O_7^{4-}$ 的化合物报道[19-21]，但其阴离子的真实结构却具有一维链状或二维层状结构，直到 1975 年，Faggiani 等人才报道了含有孤立结构二聚钒氧阴离子的化合物 $Cu_2V_2O_7$[5]。

2.1.2 结构

二钒氧簇阴离子 $V_2O_7^{4-}$ 原子连接方式如图 2-1 所示。

在二钒氧簇阴离子中，每个钒原子与 4 个氧原子相连，其中一个氧原子为共用桥氧原子，与另一个钒原子相连，其余的氧原子为非共用端基氧原子，钒原子为 4 配位四面体构型，整个二聚离子构成一个共顶点的双四面体结构（图 2-2）。结构优化结果表明，$V_2O_7^{4-}$ 应具有 D_{3h} 或 D_{3d} 对称性[4]，但在晶体结构中，由于 $V_2O_7^{4-}$ 受到阳离子的作用，其 V—O—V 键角在相当宽的范围内（117°～180°）变化[5,6]。

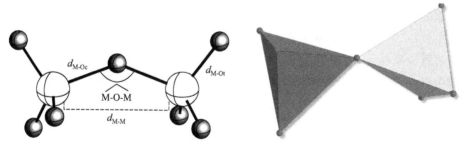

图 2-1　二钒氧簇阴离子 $V_2O_7^{4-}$ 原子连接图　　图 2-2　二钒氧簇阴离子共顶点双四面体结构

2.1.3 相关化合物

偏钒酸钠与方酸在一定条件下反应，可以得到分子组成为 $(NH_4)[V_2O_2(OH)(C_4O_4)_2(H_2O)_3]\cdot H_2O$ 的化合物[22]，其阴离子结构如图 2-3 所示。在该阴离子中，两个钒原子均为 6 配位八面体构型，每个钒原子连接两个端基氧、两个桥氧和两个桥联的方酸根，整个阴离子构成一个共棱的二聚八面体。

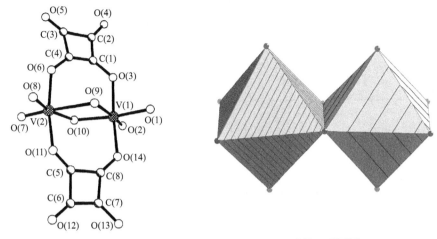

(a) $[V_2O_2(OH)(C_4O_4)_2(H_2O)_3]^-$ 原子连接图 (b) 共棱八面体结构

图 2-3　$[V_2O_2(OH)(C_4O_4)_2(H_2O)_3]^-$ 原子连接图和共棱八面体结构

2.2　二钼氧簇

2.2.1　合成

尽管分子内原子比为 $M_2Mo_2O_7$（M=K、Na、NH_4^+ 和 Ag）的化合物早已从水溶液中分离出来，但它们阴离子的真实结构并非孤立的二钼氧簇离子 $[Mo_2O_7]^{2-}$，而是一维无限链状的多聚阴离子[23-26]。1977 年，Day 和 Klemperer 报道了以 α-$[n$-$(C_4H_9)_4N]Mo_8O_{26}$ 为原料，在乙腈溶液中，用四丁基氢氧化铵碱解 α-$[n$-$(C_4H_9)_4N]Mo_8O_{26}$ 制备 $[n$-$(C_4H_9)_4N]_2Mo_2O_7$ 的方法[27]。1984 年，Braunstein 报道了用空气中的氧氧化阴离子 $[(\eta$-$C_5H_4R)Mo(CO)_3]^-$，非常意外地得到了二钼氧簇阴离子[8]。1991 年，Bhattacharyya 报道了二钼氧簇的 PPN(Ph_3P=N^+=PPh_3) 和 PPh_4 盐的合成方法[9]。具体合成步骤可见相关文献 [8,9,27,28]。

2.2.2　谱学表征

在 650～1000cm^{-1} 范围内的红外光谱分析表明，化合物 $[n$-$(C_4H_9)_4N]_2Mo_2O_7$ 在 735cm^{-1}（m，sh）、780cm^{-1}（s，br）、880cm^{-1}（s，br）、928cm^{-1}（m）和 975cm^{-1}（w）处存在振动吸收峰[28]。在化合物 $(PPN)_2[Mo_2O_7]$ 中，其特征振动峰的位置分别被认为是 $\nu_{Mo=O}$，885cm^{-1}（s）；$\nu_{(Mo\text{-}O\text{-}Mo)asym}$，795$cm^{-1}$（m）；$\nu_{(Mo\text{-}O\text{-}Mo)sym}$，450$cm^{-1}$（w）。溶液中的紫外光谱（图 2-4）表明[29]，二

钼氧簇离子在242nm附近有一弱的吸收峰,在210nm附近有一强的吸收峰,与钼酸根单体在水溶液中的吸收光谱相比(图2-5),可认为242nm附近的吸收峰为二钼氧簇阴离子$Mo_2O_7^{2-}$的特征吸收峰。其他的研究也表明二钼氧簇离子在242nm附近存在吸收峰[30,31]。

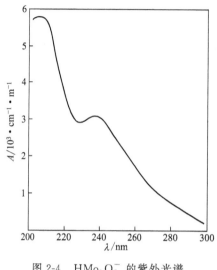

图2-4 $HMo_2O_7^-$的紫外光谱

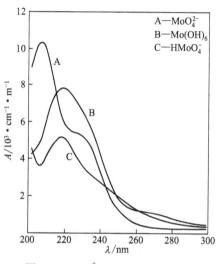

图2-5 MoO_4^{2-}、$Mo(OH)_6$和$HMoO_4^-$的紫外光谱

电喷雾质谱研究表明,二钼氧簇离子在碰撞诱导解离后,片段离子$[MoO_4]^-$和$[MoO_3]^-$的丰度基本相当,表明两个片段都是从母体离子$[O_3Mo\text{-}O\text{-}MoO_3]^{2-}$的桥氧键Mo—O断裂产生的[32]。

2.2.3 结构

$[(n\text{-}C_4H_9)_4N]_2Mo_2O_7$结构分析表明,化合物中含有2个$(n\text{-}C_4H_9)_4N^+$阳离子和1个二钼氧簇阴离子$Mo_2O_7^{2-}$。二钼氧簇离子中原子的连接方式与二钒氧簇阴离子相似(图2-1)。每个钼原子上连接4个氧原子,四面体构型,2个钼原子通过1个桥氧原子相连,整个阴离子构成共顶点的双四面体结构(图2-2)。

结构优化数据表明,二钼氧簇离子$Mo_2O_7^{2-}$倾向于采取具有直线构型的D_{3d}点群[4],但由于各构型之间的能量差很小,因此,在分子内$[Mo_2O_7]^{2-}$呈C_2[7]、C_s[8]和D_{3d}[9]构型的化合物都有报道。

2.2.4 相关化合物

1990年,Taylor等人用^{95}Mo核磁和激光拉曼光谱研究了含有HCl和HBr

的钼酸钠水溶液，发现当溶液中 HCl 和 HBr 的浓度达到 2~4mol/L 时，溶液中存在 $[Mo_2O_5(H_2O)_6]^{2+}$ [33]。分析表明，其结构应为 $[(H_2O)_3O_2MoOMoO_2(H_2O)_3]^{2+}$。

Kessler 等人将钼溶于含有 LiCl 的甲醇溶液当中，得到了分子组成为 $[LiMo_2O_2(OMe)_7(MeOH)]$ 的化合物[34]。结构分析表明（图 2-6），在该化合物中，钼原子为 6 配位八面体结构，每个钼原子连接 5 个甲醇分子和 1 个氧原子，两个钼原子通过 3 个桥联甲醇分子相连，形成 1 个共面的双八面体结构（图 2-7）。

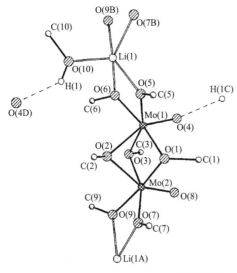

图 2-6 $[LiMo_2O_2(OMe)_7(MeOH)]$ 原子连接图

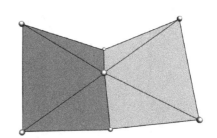

图 2-7 共面双八面体结构

2000 年，Gouzerh 等人以 $[(n\text{-}C_4H_9)_4N]_2Mo_2O_7$ 为原料，与羰基化合物反应，合成了多例具有新颖结构的金属有机衍生物[35]。

Slawin 和 Kabanos 等人将 $MoCl_5$ 溶于 1∶4 的浓盐酸的水溶液中，加入固体 $(NH_4)_2SO_3$，当加入的 $(NH_4)_2SO_3$ 使溶液的 pH 值为 6 时，得到分子组成为 $(NH_4)_8[(Mo_2O_4)(SO_3)_4(\mu_2\text{-}SO_3)] \cdot 2H_2O$ 的化合物[36]。结构分析表明（图 2-8），该化合物阴离子中钼为 6 配位，每个钼原子连接 3 个亚硫酸根和 3 个氧原子，其中 2 个氧原子和 1 个亚硫酸根作为桥联配体，两个钼原子形成共棱的双八面体结构，如图 2-9 所示。

以 $(NH_4)_6Mo_7O_{24} \cdot 4H_2O$ 和 $CoC_2O_4 \cdot 2H_2O$ 为原料，在乙二胺存在下，通过水热反应，可得到分子组成为 $NH_4(enH_2)_{0.5}[Co(en)_3][Mo_2O_7(C_2O_4)] \cdot H_2O$ 的化合物[37]，结构分析表明，该化合物阴离子为二钼氧簇（图 2-10）。

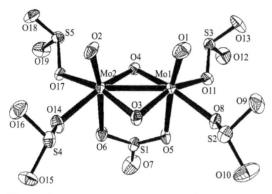

图 2-8 $[(Mo_2O_4)(SO_3)_4(\mu_2\text{-}SO_3)]^{8-}$ 原子连接图

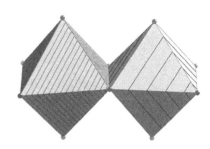

图 2-9 共棱双八面体结构

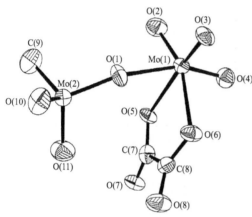

图 2-10 $[Mo_2O_7(C_2O_4)]^{4-}$ 原子结构

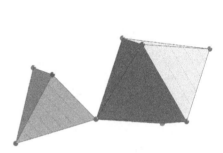

图 2-11 四面体和八面体共顶点结构

有意思的是两个钼原子的配位环境不同，1个钼原子与4个氧原子配位，为四面体构型，另1个钼原子与4个氧原子和草酸根中的两个氧原子配位，为八面体构型，两个钼原子通过1个桥联氧原子相连，整个阴离子由共顶点的四面体和八面体构成（图2-11）。

2003年，王恩波等人以 $(NH_4)_6Mo_7O_{24}\cdot 4H_2O$ 和邻菲啰啉为原料，在 SeO_2 存在下，在水热170℃条件下，合成了分子组成为 $Mo_2O_5(ophen)_2$ 的化合物[38]，结构如图2-12所示。

分子中两个钼原子均为6配位，八面体构型。该化合物在生成时，伴随有邻菲啰啉的氧化反应，在 SeO_2 的催化下，邻菲啰啉被氧化为2-羟基邻菲啰啉，2-羟基邻菲啰啉再和钼配位。每个钼原子与1个2-羟基邻菲啰啉中的两个氮原子和另1个2-羟基邻菲啰啉羟基脱去质子后的氧原子，1个桥氧原子和2个端氧原子相连，两个钼原子构成共顶点的二聚八面体结构（图2-13）。

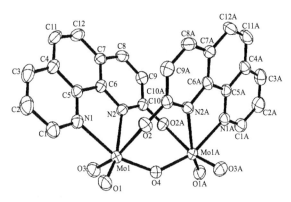

图 2-12 $Mo_2O_5(ophen)_2$ 原子连接图

图 2-13 共顶点二聚八面体结构

参考文献

[1] Mestres J, Duran M, Martin-Zarza P, Medina de la Rosa E, Gili P. Ab initio theoretical study on geometries, chemical bonding, and infrared and electronic spectra of the $M_2O_7^{2-}$ (M=Cr, Mo, W) anions. Inorg Chem, 1993, 32: 4708-4713.

[2] Ribeiro-Claro P J A, Amado A M, Teixeira-Dias J J C. Structures and vibrational frequencies of vanadium(V) oligomers: an ab initio study using effective core potentials. J Comput Chem, 1996, 17: 1183.

[3] Amado A M, Ribeiro-Claro P J A. Ab initio calculations on some transition metal heptoxides by using effective core potentials. J Mol Struct Theochem, 1999, 469: 191.

[4] Bridgeman A J, Cavigliasso G. Structure and bonding in dinuclear oxoanions of V, Nb, Ta, Mo, and W. J Phys Chem A, 2001, 105: 7111-7117.

[5] Calvo C, Faggiani R. Cupric divanadate. Acta Crystallogr B, 1975, 31: 603.

[6] Tytko K H, Mehmke J, Fischer S. Bonding and charge distribution in isopolyoxometalate ions and relevant oxides——A bond valence approach. Struct Bonding, 1999, 93: 129.

[7] Day V W, Fredrich M F, Klemperer W G, Shum W. Synthesis and characterization of the dimolybdate ion, $Mo_2O_7^{2-}$. J Am Chem Soc, 1977, 99 (18): 6146-6148.

[8] Braunstein P, De Meric de Bellefon C, Lanfranchi M, Tiripicchio A. First heterobimetallic complexes with bridging and chelating $Ph_2PCH_2PPh_2$ (dppm): crystal structure of $[(\eta\text{-}C_5H_4Me)Mo(CO)_2(\mu\text{-dppm})Pt(dppm)]_2[Mo_2O_7]$. Air oxidation of the anions $[(\eta\text{-}C_5H_4R)Mo(CO)_3]^-$ into $[Mo_2O_7]^{2-}$. Organometallics, 1984, 3 (11): 1772-1774.

[9] Bhattacharyya R, Biswas G S. Oxygen transfer from peroxometalates as a new and general route to the synthesis of oxopolymetalates: rational synthesis of $Mo_2O_7^{2-}$, $Mo_6O_{19}^{2-}$ and $Mo_7O_{24}^{6-}$. Evidence of a $M_2O_7^{2-}$ with linear M—O—M axis. Inorganica Chimica Acta, 1991, 181 (2): 213-216.

[10] Errington R J, Kerlogue M D, Richards D. G. Non-aqueous routes to a new polyoxo-

tungstate. Chem Commun, 1993 (7): 649.

[11] Clegg W, Errington R J, Fraser K A, Richards D G. Evidence for rapid ligand redistribution in non-aqueous tungstate chemistry: rational synthesis of the binuclear tungsten oxoalkoxide $[W_2O_5(OMe)_4]^{2-}$. Chem Commun, 1993, 13 (13): 1105-1107.

[12] Zhai H J, Huang X, Waters T, Wang X B, O'Hair R A J, Wedd A G, Wang L S. Photoelectron spectroscopy of doubly and singly charged group VIB dimetalate anions: $M_2O_7^{2-}$, $MM'O_7^{2-}$, and $M_2O_7^{-}$ (M, M'=Cr, Mo, W). J Phys Chem A, 2005, 109: 10512-10520.

[13] Jander G, Jahr K F. Über amphotere oxydhydrate, deren wäßrige lösungen und kristallisierende verbindungen. XVII. mitteilung. Aufbau und abbau höhermolekularer, anorganischer verbindungen in lösung am beispiel der vanadinsäuren, polyvanadate und vanadansalze. Z Anorg Chem, 1933, 212 (1): 1-20.

[14] Souchay P, Carpeni G. Sur la constitution des vanadates en solution aqueuse. Bull Soc Chim, 1946: 160.

[15] Lelong M. Étude sur le système H_2O-V_2O_5-Na_2O (∗). Bevue de Chimie minérale, 1966: 259.

[16] Tadros T F, Sadek H, El-Harakani A A. Studies on acidified alkali vanadate solutions. Z phys Chem, 2017, 242O (1): 1-17.

[17] Long G G, Stanfield R L, Hentz Jr F C. Comparison of strong acid and weak acid tltration curves. Journal of Chemical Education, 1979, 56 (3): 194-195.

[18] Prasad S. Electrometric studies on the isopolyanions of Vanadium (V). An Acad Brasil Ciênc, 1981, 53 (3): 471.

[19] Schindler M, Hawthorne F C. Structural characterization of the β-$Cu_2V_2O_7$-α-$Zn_2V_2O_7$ solid solution. J Solid State Chem, 1999, 146: 271-276.

[20] Obbade S, Dion C, Saadi M, Abraham F. Synthesis, crystal structure and electrical characterization of $Cs_4[(UO_2)_2(V_2O_7)O_2]$, a uranyl divanadate with chains of corner-sharing uranyl square bipyramids. J Solid State Chem, 2004, 177: 1567-1574.

[21] Xiao D R, Wang S T, Hou Y, Wang E B, Li Y G, An H Y, Xu L, Hu C W. Hydrothermal synthesis an crystal structure of a new layered titanium vanadate decorated with organonitrogen ligand: $[Ti(2,2'-bpy)V_2O_7]$. J Mol Struct, 2004, 692, 107-114.

[22] Khan M I, Chang Y D, Chen Q, Salta J, Lee Y S, O'Connor C J, Zbieta J. Synthesis and characterization of binuclear oxo-vanadium complexes of carbon oxoanion ligands. Crystal structure of the binuclear vanadium (IV) complex $(NH_4)[V_2O_2(OH)(C_2O_4)_2(H_2O)_3] \cdot H_2O$, of the mixed-valence vanadium (V)/vanadium (IV)-squarate species $[(n-C_4H_9)_4N][V_2O_3(C_4O_4)_2(H_2O)_3] \cdot 3H_2O$ and $[(C_4H_9)_4N]_4[V_4O_6(C_4O_4)_5(H_2O)_4] \cdot 6H_2O$, and of the binuclear vanadium (IV)-oxalate species

$[V_2O_2Cl_2(C_2O_4)(CH_3OH)_4] \cdot 2Ph_4Cl$. Inorg Chem, 1994, 33: 6340-6350.

[23] Gatehouse B M, Jozsa A J. The crystal structure of potassium dimolybdate hydrate. J Solid State Chem, 1987, 71: 34-39.

[24] Seleborg M. A refinement of the crystal structure of disodium dimolybdate. Acta Chem Scand, 1967, 21: 499-504.

[25] Knopnadel I, Hartl H, Hunnius W D, Fuchs J. Structure of diimine: X-ray diffraction analysis of $N_2H_2[Cr(CO)_5]_2 \cdot 2THF$. Angew Chem Int Ed Engl, 1974, 13: 823.

[26] Armour A W, Drew M G B, Mitchell P C H. Crystal structures of silver dimolybdate, $Ag_2Mo_2O_7$, and silver ditungstate, $Ag_2W_2O_7$. J Chem Soc, Dalton Trans, 1976: 1316.

[27] Day V W, Fredrich M F, Klemperer W G, Shum W. Synthesis and characterization of the dimolybdate ion, $Mo_2O_7^{2-}$. J Am Chem Soc, 1977, 99 (18): 6146.

[28] Ginsberg A P. Inorganic Synthesis: V27. New York: John Wiley & Sons. Inc, 1990: 79.

[29] Cruywagen J J, Heyns J B B. Equilibria and UV spectra of mono-and polynuclear molybdenum (Ⅵ) species. Inorg Chem, 1987, 26: 2569-2572.

[30] Cruywagen J J, Heyns J B, Rohwer E F C H. Dimeric cations of molybdenum (Ⅵ). J Inorg and Nucl Chem, 1978, 40 (1): 53-59.

[31] Cruywagen J J, Heyns J B, Water R F van de. A potentiometric, spectrophotometric, and calorimetric investigation of molybdenum (Ⅵ)-oxalate complex formation. Cheminform, 1987, 18 (4): 1857-1862.

[32] Lau T C, Wang J Y, Guevremont R, Siu K W M. Electrospray tandem mass spectrometry of polyoxoanions. J Chem Soc Chem Commun, 1995: 877-878.

[33] Coddington J M, Taylor M J. Molybdenum-95 nuclear magnetic resonance and vibrational spectroscopic studies of molybdenum (Ⅵ) species in aqueous solutions and solvent extracts from hydrochloric and hydrobromic acid: evidence for the complexes $[Mo_2O_5(H_2O)_6]^{2+}$, $[MoO_2X_2(H_2O)_2]$ (X=Cl or Br), and $[MoO_2Cl_4]^{2-}$. J Chem Soc Dalton Trans, 1990 (1): 41-47.

[34] Kessler V G, Panov A N, Turova N Y, Starikova Z A, Yanovsky A I, Dolgushin F M, Pisarevsky A P, (the late) Struchkov Y T. Anodic oxidation of molybdenum and tungsten in alcohols: isolation and X-ray single-crystal study of side products. J Chem Soc Dalton Trans, 1998, 267 (1): 21-30.

[35] Villanneau R, Delmont R, Proust A, Gouzerh P. Merging organometallic chemistry with polyoxometalate chemistry. chem Eur J, 2000, 6 (7): 1184-1192.

[36] Manos M J, Woollins J D, Slawin A M Z, Kabanos T A. Polyoxomolybdenum (Ⅴ) sulfite complexes: synthesis, structural, and physical studies. Angew Chem Int Ed, 2002, 41 (15): 2801-2805.

[37] Lin B Z, Liu P D. Hydrothermal synthesis and characterization of an asymmetric binucle-

ar molybdenum complex with oxalate ligand, $NH_4(enH_2)_{0.5}[Co(en)_3][Mo_2O_7(C_2O_4)]\cdot H_2O$. J Mol Struct, 2003, 654: 55-60.

[38] Xiao D R, Hou Y, Wang E B, Wang S T, Li Y G, De G J H, Xu L, Hu C W. Hydrothermal synthesis and crystal structure of a novel polyoxomalybdate with the hydroxylated *N*-heterocycle ligand: $Mo_2O_5(ophen)_2$ (Hophen=2-hydroxy-1, 10-phenanthroline). J Mol Struct, 2003, 654: 13-21.

第3章 三金属氧簇

由三金属形成的阴离子簇,应存在开环和闭环两种空间构型。对以六配位八面体结构为常见构型的钼、钨原子来说,开环构型将导致多个配位氧原子处于端基位置,过多的端基氧将导致阴离子簇内键价不平衡,因而难以稳定存在;而闭环构型,不仅存在端基氧过多,键价不平衡的问题,并且,由于键角不匹配导致的化学键张力增大,阴离子簇也很难形成。因此,就目前文献看,尽管阴离子组成可写作钼、钨三金属簇的化合物有很多[1-16],但配体仅为氧原子的化合物均为多聚物,而具有孤立结构的钼、钨三金属簇尚未见报道,只有当含氧有机配体参与形成化合物时,由于氧的键价可以通过与碳相连得到满足,才有可能形成孤立结构的钼、钨三金属簇[17-19]。对于钒形成的三金属簇,由于钒作中心原子的特征结构为四配位四面体构型,不论是开环构型还是闭环构型,都不存在端基氧过多或键角张力太大的问题,因此,分子组成为三钒氧簇的多聚化合物有很多[20-27],但也存在分立结构的三钒氧簇[28-30]。

3.1 三钒氧簇

3.1.1 合成

由于水溶液体系中多种聚钒氧离子共存且可相互转化[31],使得对其中某一组分的确定难度较大。早在 1959 年就有在溶液中存在三钒氧簇阴离子 $[V_3O_x]^{y-}$ 的报道[32]，但该阴离子的存在一直没有形成广泛的共识。支持开环三钒氧簇阴离子 $[V_3O_{10}]^{5-}$ 和闭环三钒簇氧阴离子 $[V_3O_9]^{3-}$ 存在的文献有很多，通过大量的电位分析和 ^{51}V 核磁实验证明了该类阴离子的存在[32-40]，但也有大量的文献用类似的方法证明三钒氧簇离子不存在[41-43]。

1985 年，Kato 等人报道了阴离子具有弯曲结构的化合物 $K_5V_3O_{10}$ 和阴离子具有线型结构的 $Na_5V_3O_{10}$ [28,29] 等含有开环型阴离子的化合物晶体结构，确切地证明了孤立的具有开环结构的三钒氧簇离子的存在。2001 年，Wilker 分离出了由闭环结构的三钒氧簇阴离子构成的化合物 $[(C_4H_9)_4N]_3(V_3O_9)$ [30]，由此确定了闭环型三钒氧簇阴离子的存在。

3.1.2 谱学表征

Wilker 等人对 $[(C_4H_9)_4N]_3(V_3O_9)$ 的 ^{51}V 核磁谱进行了系统的研究[30]，$[(C_4H_9)_4N]_3(V_3O_9)$ 在 25℃时浓度为 50mmol/L 的 CD_3CN 溶液中的 ^{51}V 核磁谱，在 -569 和 -576 出现两个共振峰（图 3-1 曲线 A），考虑到在闭环阴离子 $[V_3O_9]^{3-}$ 的晶体结构中所有的 V 原子处于相同的环境中，只应该出现一个共振峰。将样品冷却到 -10℃，则两个峰变宽，而且峰的比例发生了变化，温度恢复到 25℃，共振峰恢复原来的位置和比例。进一步将样品升温到 50℃，两个峰均变尖且移向 -569（图 3-1 曲线 B）。这些结果表明溶液中存在两个物种的相互平衡。化合物在 CD_3CN 溶液中的浓度也对 ^{51}V 核磁谱产生影响（图 3-2）。

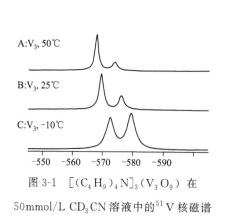

图 3-1 $[(C_4H_9)_4N]_3(V_3O_9)$ 在 50mmol/L CD_3CN 溶液中的 ^{51}V 核磁谱

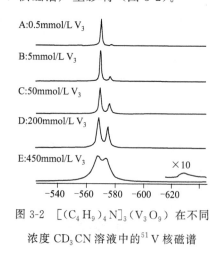

图 3-2 $[(C_4H_9)_4N]_3(V_3O_9)$ 在不同浓度 CD_3CN 溶液中的 ^{51}V 核磁谱

当溶液中 $[V_3O_9]^{3-}$ 的浓度由 0.5mmol/L 逐渐增加到 450mmol/L 时，谱图中逐步出现第二个峰且共振峰逐渐变宽。为了探明产生第二个峰的原因，

Wilker 等人考察了 $[(C_4H_9)_4N]_3[V_3O_9]$ 与 $[(C_4H_9)_4N]_3[HV_4O_{12}]$、$[(C_4H_9)_4N]_3[V_5O_{14}]$ 与 $[(C_4H_9)_4N]_3[H_3V_{10}O_{28}]$ 分别混合后的乙腈溶液的 ^{51}V 核磁共振谱（图 3-3），发现化合物 $[(C_4H_9)_4N]_3[V_3O_9]$ 溶液 ^{51}V 核磁共振谱中的第二个峰的出现与上述物种无关[30]。该峰的出现是否和另一个三聚钒氧阴离子 $[V_3O_{10}]^{5-}$ 相关，还需要进一步的实验证明。

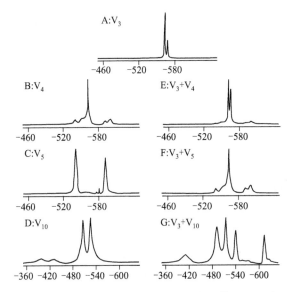

图 3-3 不同钒酸盐在 CD_3CN 溶液中的 ^{51}V 核磁谱

V_3—$[(C_4H_9)_4N]_3[V_3O_9]$；V_4—$[(C_4H_9)_4N]_3[HV_4O_{12}]$；

V_5—$[(C_4H_9)_4N]_3[V_5O_{14}]$；$V_{10}$—$[(C_4H_9)_4N]_3[H_3V_{10}O_{28}]$；

化合物的真实浓度均为 50mmol/L

3.1.3 结构

三钒氧簇阴离子存在开环和闭环两种结构类型，其组成分别为 $[V_3O_{10}]^{5-}$ 和 $[V_3O_9]^{3-}$。组成为 $[V_3O_{10}]^{5-}$ 的阴离子为开环结构，存在角型和线型两种构型。图 3-4 为开环结构阴离子 $[V_3O_{10}]^{5-}$ 的角型结构图[图 3-4(a)]和线型结构图[图 3-4(b)]。由图中可以看出，不论是线型还是角型，阴离子中的钒原子均为四配位的四面体构型，按照所处环境的不同，钒原子可分为两类：一类连接三个端氧原子和一个桥氧原子，位于链的两端；另一类连接两个端氧原子和两个桥氧原子，位于钒氧链的中间。氧原子也可分为两类：只与一个钒原子相连的为端基氧；连接两个钒原子的为桥氧。角型结构阴离子中，端氧键 V—O 平均键长为 1.648Å（1Å=10^{-10}m），桥氧键 V—O 平均键长为 1.788Å，线型分子中端氧键 V—O 平均键长为 1.669Å，桥氧键 V—O 平均键长为 1.836Å，O—V—O 键角介于 105.2°到 114.4°之间。二者位于中心位置钒上的 V—O 键角没有明显差别，其差别仅在于构象不同，这可能是由于二者的配位环境不同所造成的。

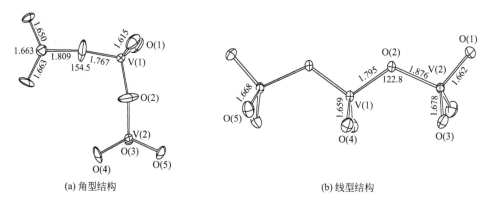

(a) 角型结构　　　　　　　　　　(b) 线型结构

图 3-4　三钒氧簇阴离子 $[V_3O_{10}]^{5-}$ 原子连接图

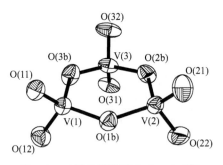

图 3-5　三钒氧簇阴离子 $[V_3O_9]^{3-}$ 原子连接图

组成为 $[V_3O_9]^{3-}$ 的三聚钒氧簇离子为闭环结构。图 3-5 为闭环结构阴离子 $[V_3O_9]^{3-}$ 的结构，由图可知，三个钒原子处于相同的环境中，每个钒原子连接两个端氧和两个桥氧，均为四配位四面体构型，三个钒原子通过三个桥氧原子连接成闭合的三元环，三元环具有畸变的船式构型，其中的氧原子可分为桥氧与端氧两类，桥氧键 V—O 平均键长为 1.80Å，端氧键 V—O 平均键长为 1.63Å。

3.2　三钼钨氧簇

前已述及，由于钼原子和钨原子的空间构型的要求，仅以氧原子参与形成的三钼氧簇和三钨氧簇阴离子很难形成，因此，到目前为止，尚未有仅以氧原子形成的三钼氧簇和三钨氧簇阴离子构成化合物的报道。但当含氧有机配体参与化合物的形成时，由于氧上键价可通过与其他原子的键合来满足，可以形成含有三个钼原子的聚金属氧簇化合物。1989 年，Zubieta 等人报道了几例该类化合物的合成和结构[17-19]。

参考文献

[1] Gatehouse B M, Leverett P. The crystal structure of dipotassium trimolybdate, $K_2Mo_3O_{10}$; a

compound with five-co-ordinate molybdenum (Ⅵ). Chem Commun, 1968, 6 (8): 374-375.

[2] Guillou N, Férey G. Hydrothermal synthesis and crystal structure of anhydrous ethylenediamine trimolybdate ($C_2H_{10}N_2$)[Mo_3O_{10}]. J Solid State Chem, 1997, 132: 224-227.

[3] Lasocha W, Jansen J, Schenk H. Crystal structure of fibrillar anilinum trimolybdate $2(C_6H_5NH_3) \cdot Mo_3O_{10} \cdot 4H_2O$ from X-ray powder data. J Solid State Chem, 1995, 117: 103-107.

[4] Lasocha W, Jansen J, Schenk H. Crystal structure of ammonium trimolybdate monohydrate $(NH_4)_2Mo_3O_{10} \cdot H_2O$ by powder diffraction method. J Solid State Chem, 1995, 116: 422-426.

[5] Lasocha W, Jansen J, Schenk H. Application of a new method for the determination of accurate intensities from powder diffraction data. Crystal structure determination of fibrillar silver trimolybdate. J Solid State Chem, 1994, 109: 1-4.

[6] Lasocha W, Jansen J, Schenk H. Crystal structure of fibrillar potassium trimolybdate $K_2Mo_2O_{10} \cdot 3H_2O$ by direct method/powder diffraction package. J Solid State Chem, 1995, 115: 225-228.

[7] Enjalbert R, Guinneton F, Galy J. $Cs_2Mo_3O_{10}$. Acta Cryst, 1999, C55: 273-276.

[8] Range K J, Fässler A. Diammonium trimolybdate (Ⅵ), $(NH_4)_2Mo_3O_{10}$. Acta Cryst, 1990, C46: 488-489.

[9] Range K J, Bauer K, Koement U. $(NH_4)Mo_3O_9$, an ammonium molybdenum bronze. Acta Cryst, 1990, C46: 2007-2009.

[10] Forster A, Kreusler H U, Fuchs J. Die kristallinen phasen der alkalitrimolybdate. Z Naturforsh, 1985, 40b: 1139-1148.

[11] Kreusler H U, Förster A, Fuchs J. Die struktur des rubidiumtrimolybdathydrats $Rb_2Mo_3O_{10} \cdot H_2O$. Z Naturforsch, 1980, 35b: 242-244.

[12] Selerorg M. The crystal structure of dipotassium trimolybdate. Acta Chem Scand, 1966, 20 (8): 2195-2201.

[13] Nomiya K. Topology and chemistry of polyoxometalates: inorganic polymers formed by connecting octahedral units, hetero-and isopolyanions, and their strucrural stability. Polyhedron, 1987, 6 (2): 309-314.

[14] Khan M I, Chen Q, Zubieta J. Hydrothermal synthesis and crystal structure of the trimolybdate, $(H_3NCH_2CH_2NH_3)Mo_3O_{10}$. Inorg Chim Acta, 1993, 213: 325-327.

[15] Hodorowicz S A, Lasocha W. Structural characterization of the ammonium trimolybdate hydrate $(NH_4)_2Mo_3O_{10} \cdot H_2O$. Cryst Res Technol, 1988, 23 (2): k43-k46.

[16] Yan B B, Xu Y, Goh N K, et al. Hydrothermal synthesis and crystal structure of a novel one-dimensional tritungstate $(C_2H_{10}N_2)[W_3O_{10}]$. Inorg Chem Commun, 2000, 3: 379-382.

[17] Ma L, Liu S C, Zubieta J. Synthesis and characterization of a trinuclear polyoxomolybdate containing a reactive [MoO_3] unit, [(n-C_4H_9)$_4$N]$_2$[$Mo_3O_7(CH_3C(CH_2O)_3)_2$],

and its conversion to the methoxy derivative $[(n\text{-}C_4H_9)_4N][Mo_3O_6(OCH_3)(CH_3C(CH_2O)_3)_2]$. Inorg Chem, 1989, 28: 175-177.

[18] Gumaer E, Lettko K, Ma L, et al. Trinuclear polyoxomolybdates, the crystal and molecular structure of $[(n\text{-}C_4H_9)_4N][Mo_3O_6(OCH_2CH_2Cl)\{(OCH_2)_3CCH_3\}_2]$. Inorganica Chimica Acta, 1991, 179 (1): 47-51.

[19] Chen Q, Ma L, Liu S C, et al. Structual characterization of the pentamolybdate anion, $[(MoO_4)_2\{Mo_3O_8(OMe)\}]^{3-}$, and isolation of the $[Mo_3O_8(OMe)]^+$ trinuclear core in the squarate complex $[\{Mo_3O_8(OMe)\}(C_4O_4)_2]^{3-}$. J Am Chem Soc, 1989, 111: 5944-5946.

[20] Evans H T, Jr Block S. The crystal structures of potassium and cesium trivanadates. Inorg Chem, 1966, 5 (10): 1808-1814.

[21] Benchrifa R, Leblanc M, De Pape R. Structure of the trivanadate TlV_3O_8. Acta Cryst, 1990, C46: 177-179.

[22] Block S. Crystal structure of potassium trivanadate. Nature, 1960, 186: 540-541.

[23] Lin B Z, Liu S X. Ammonium trivanadate (V), $NH_4V_3O_8$. Acta Cryst, 1999, C55: 1961-1963.

[24] Ammari L E, Azrour M, Depmeier W, et al. Cadmium sodium trivanadate. Acta Cryst, 1995, C55: 1743-1746.

[25] Rettich R, Müller-Buschbaum H. Darstellung und strukturbeschreibung von $NaCa_3Mn(V_2O_7)(V_3O_{10})$. Z Naturforsch, 1998, 53b: 507-511.

[26] Chirayil T G, Boylan E A, Mamak M, et al. $NMe_4V_3O_7$: critical role of pH in hydrothermal synthesis of vanadium oxides. Chem Commun, 1997, 33-34.

[27] 李亚丰, 崔巍, 朱广山, 施展, 许宪祝, 裘式纶. $\alpha\text{-}(+)\text{-}Co(en)_3V_3O_9 \cdot H_2O$ 的合成与表征. 高等学校化学学报, 2002, 23 (7): 1243-1245.

[28] Kato K, Takayama-Muromachi E. Pentasodium trivanadate dihydrate. Acta Cryst, 1985, C41: 1409-1411.

[29] Kato K, Takayama-Muromachi E. Pentapotassium Trivanadate, $K_5V_3O_{10}$. Acta Cryst, 1985, C41: 647-649.

[30] Hamilton E E, Fanwick P E, Wilker J J. The elusive vanadate $(V_3O_9)^{3-}$: isolation, crystal structure, and nonaqueous solution behavior. J Am Chem Soc, 2002, 124 (1): 78-82.

[31] Souchay P, Carpeni G. Sur la constitution des vanadates en solution aqueuse. Bull Soc Chim, 1946: 160.

[32] Ingri N, Brito F. Equilibrium studies of polyanions. VI. polyvanadates in alkaline Na(Cl) medium. Acta Chem Scand, 1959, 13: 1971-1996.

[33] Hatton J V, Saito Y, Schneider W G. Nuclear magnetic resonance investigations of some group V metal fluorides and oxyions. Can J Chem, 1965, 43: 47-56.

[34] Howarth O W, Richards R E. Nuclear magnetic resonance study of polyvanadate equilib-

[35] Pope M T, Dale B W Q. Isopoly-vanadates, -niobates, and otantalates. Rev Chem Soc, 1968, 22: 527-548.

[36] Kepert D L. Comprehensive inorganic chemistry//Compounds of the transition elements involving metal-metal bonds. New York: Pergamon Press, 1973: 197-228.

[37] Habayeb M A, Hileman O E Jr. ^{51}V FT-NMR investigations of metavanadate ions in aqueous solutions. Can J Chem, 1980, 58: 2255-2261.

[38] Tracey A S, Jaswal J S, Angus-Dunne S J. Influences of pH and ionic strength on aqueous vanadate equilibria. Inorg Chem, 1995, 34 (22): 5680-5685.

[39] Andersson I, Pettersson L, Hastings J J, et al. Oxygen and vanadium exchange processes in linear vanadate oligomers. J Chem Soc Dalton Trans, 1996: 3357-3361.

[40] Heath E, Howarth O W. Vanadium-51 and oxygen-17 nuclear magnetic resonance study of vanadate (V) equilibria and kinetics. J Chem Soc Dalton Trans, 1981: 1105-1110.

[41] Pettersson L, Hedman B, Andersson I, et al. Multicomponent polyanions. A potentiometric and ^{51}V NMR study of equilibria in the H^+-HVO_4^{2-} system in 0.6M Na(Cl) medium covering the range $1<-\lg[H^+]<10$. Chem Scr, 1983. 22: 254-264.

[42] O'Donnell S E, Pope M T. Applications of vanadium-51 and phosphorus-31 nuclear magnetic resonance spectroscopy to the study of iso-and hetero-polyvanadates. J Chem Soc Dalton Trans, 1976: 2290-2297.

[43] Pope M T. Comprehensive Coordination Chemistry: The synthesis, reactions, properties & applications of coordination compounds ligands. UK: Pergamon Press, 1987: 1179.

第4章 四金属氧簇

四金属氧簇与二金属氧簇和三金属氧簇相比，由于阴离子内金属原子个数增加，其形成阴离子簇的结构应该更具多样性。对于由钒形成的具有孤立结构的四金属氧簇阴离子来说，由于原子的四面体构型要求，四钒氧簇阴离子应该具有环状和链状两种构型。而以六配位八面体构型为主的钼、钨原子来说，无论是形成环状结构还是链状结构，阴离子簇中都需要有大量的氧原子参与，仅有氧原子作为配原子形成的钼、钨四金属氧簇很难实现价态和电荷的平衡，因此，钼、钨四金属氧簇的形成必须由含氧有机配体以氧原子参与配位才能够完成，以满足中心原子对空间构型的要求。

4.1 四钒氧簇

4.1.1 合成

1976 年，Fuchs 报道由五氧化二钒（V_2O_5）与四丁基氢氧化铵的乙醇溶液反应，得到了第一个含有孤立结构的四钒氧簇阴离子的化合物[$(C_4H_9)_4N$]$_3HV_4O_{12}$，这是见于报道的第一个四钒氧簇阴离子[1]。1986 年，Millet 报道了一个四钒氧簇化合物[2]，1987 年，Gatehouse 等人确定该化合物阴离子为链状结构的四钒氧簇[3]。杨国昱等人在1998年报道了由水热法合成的分子组成为 $[Ni(C_{10}H_8N_2)_3]_2[V_4O_{12}] \cdot 11H_2O$ 的化合物结构[4]。Yagasaki 等人将 V_2O_5（2g）溶于10%的四乙基氢氧化铵的水溶液中（35mL），得到分子组成为 $[C_8H_{20}N]_4[V_4O_{12}] \cdot 2H_2O$ 的化合物[5]。

除上述几例以外，到目前为止已报道的含有孤立结构聚四钒氧簇阴离子的化合物还有 $[Ni(phen)_3]_2[V_4O_{12}] \cdot 7H_2O$、$[Ni(phen)_3][V_4O_{12}] \cdot 17.5H_2O$、$[Fe(C_{10}H_8N_2)_3]_2[V_4O_{12}] \cdot 10H_2O$、$[Zn(bpy)_3]_2[V_4O_{12}] \cdot 11H_2O$[6-9] 等。

4.1.2 谱学表征

$[Ni(phen)_3]_2[V_4O_{12}] \cdot 7H_2O$ 的红外光谱中，出现如下振动吸收峰：$938cm^{-1}$（s）、$898cm^{-1}$（s）、$871cm^{-1}$（w*）、$853cm^{-1}$（m）、$815cm^{-1}$（sh）、$795cm^{-1}$（vs）、$777cm^{-1}$（s）、$724cm^{-1}$（s*）、$645cm^{-1}$（w）、$512cm^{-1}$（w）、$427cm^{-1}$（w*），其中带有*标记的为邻菲啰啉振动峰，$938cm^{-1}$ 和 $898cm^{-1}$ 的峰可归属为 V—O_t（端基氧）的振动吸收，位于 $815cm^{-1}$、$795cm^{-1}$、$777cm^{-1}$、$645cm^{-1}$ 的峰可归属于 V—O_b（桥氧）的振动吸收[6]。

4.1.3 结构

四钒氧簇阴离子存在环状和链状两种结构，由于链状结构阴离子中位于两端的钒原子上分别有三个端基氧原子，因此，环状结构四钒氧簇阴离子与链状结构四钒氧簇阴离子组成具有明显差别，分别为 $[V_4O_{12}]^{4-}$ 和 $[V_4O_{13}]^{6-}$。从现有文献看，四钒氧簇阴离子具有环状结构的化合物较多。环状阴离子结构如图4-1

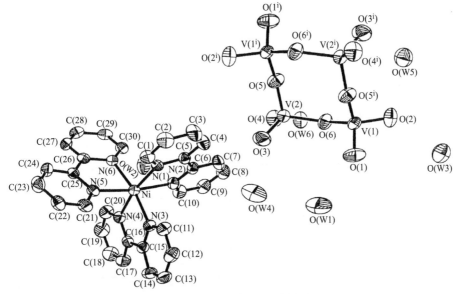

图 4-1 $[V_4O_{12}]^{4-}$ 原子连接图

图 4-2 $[V_4O_{12}]^{4-}$ 共顶点四面体

所示。4 个钒原子通过 4 个桥氧相连构成一个四元环，每个钒原子上又连接两个端氧原子，四个桥氧原子位于一个平面上，四个钒氧原子交替地偏离平面。钒原子为四面体构型，整个阴离子为一个由 4 个共顶点相连的四面体构成的环（图 4-2）。阴离子中所有钒原子的环境相同，氧原子分为端基氧和桥氧两类。不同化合物中的 V—O 键长列于表 4-1。

表 4-1 化合物中的 V—O 键长　　　　　　　　　　　　　单位：Å

	[Ni(phen)]$_3$ [V$_4$O$_{12}$]· 17.5H$_2$O	[Ni(phen)$_3$]$_2$ [V$_4$O$_{12}$]· 7H$_2$O	[Zn(bpy)$_3$] [V$_4$O$_{12}$]· 11H$_2$O	[Ni(C$_{10}$H$_8$N$_2$)$_3$]$_2$ [V$_4$O$_{12}$]· 11H$_2$O	Ba$_3$V$_4$O$_{13}$
V—O$_t$	1.625, 1.635, 1.622	1.618, 1.647, 1.628, 1.648	1.631, 1.629, 1.649	1.621, 1.638, 1.641	1.672, 1.688(链端) 1.646, 1658(链中)
V—O$_b$	1.765, 1.774, 1.778, 1.784	1.791, 1.794, 1.770, 1.784	1.804, 1.789, 1.777, 1.793	1.790, 1.803, 1.782, 1.784	1.820(链端) 1.773, 1.800(链中)

具有链状结构四钒氧簇阴离子的化合物目前报道的仅有 Ba$_3$V$_4$O$_{13}$ 一例[2]。在该化合物中，阴离子具有链状结构（图 4-3），阴离子链像一个扭曲的 V 形。在该阴离子中，根据配位环境的不同，钒原子分为两类：位于链两端的钒原子，分别连接三个端基氧和一个桥氧；位于链中间的两个钒原子，分别连接两个端基氧和两个桥氧，键长列于表 4-1。

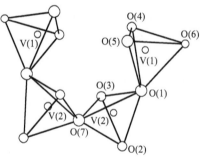

图 4-3 $[V_4O_{13}]^{6-}$ 结构

4.1.4 相关化合物

前面介绍了由四钒氧簇阴离子与抗衡阳离子（配离子或有机胺离子）形成的化合物，在那些化合物中，四钒氧簇阴离子与阳离子之间没有形成化学键，而只是靠静电引力相互作用。由于四钒氧簇阴离子中的每一个钒原子上都有两个端基氧原子，因此，很容易与能提供空轨道的金属离子发生配位作用，形成与金属配离子配位的四钒氧簇化合物。

1990 年，Klemperer 等人报道了含有四钒氧簇阴离子基团的化合物 $[(\eta\text{-}C_8H_{12})Ir]_2(V_4O_{12})$[10]，该化合物中的四钒氧簇阴离子基团与孤立结构中的阴离子簇具有相同的结构特征，其不同之处仅在于同 $[(\eta\text{-}C_8H_{12})Ir]^+$ 基团发生了共价相连。在后来的研究工作中，又报道了[{(η^3-C_4H_7)Pd}$_2$(V_4O_{12})]$^{2-[11]}$、[{Co(phen)$_2$}$_2V_4O_{12}$]·H$_2$O$^{[12,13]}$、[{Cd(bpy)$_2$}V_4O_{12}]$^{[14]}$、[{Cd(phen)$_2$}V_4O_{12}]·5H$_2$O$^{[7]}$、[{Zn(bpy)$_2$}$_2V_4O_{12}$]、[{Zn(phen)$_2$}V_4O_{12}]·H$_2$O$^{[9]}$ 和 [{Ni(quaterpy)(H$_2$O)}$_2V_4O_{12}$]·10H$_2$O$^{[15]}$ (quaterpy=2,2′;2″;6″,2‴四联吡啶) 等化合物的合成和结构。在这些化合物中，四钒氧簇阴离子与金属离子的配位方式分为两种：一种是每个钒原子上由一个端基氧与金属离子配位，每个金属离子与两个钒原子上的端基氧配位，其他空轨道由有机配体配位，图 4-4 为该类配位方式；另一种配位方式为金属离子与一个钒原子上的一个端氧配位，两个配位金属离子相互处于间位与聚四钒氧簇阴离子配位，金属离子上的其他空间轨道由有机配体和水配位，化合物 [{Ni(quaterpy)(H$_2$O)}$_2V_4O_{12}$]·10H$_2$O 即只有这种配位方式（图 4-5）。

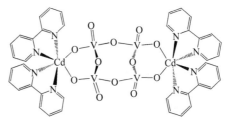

图 4-4　[{Cd(bpy)$_2$} V$_4$O$_{12}$]结构

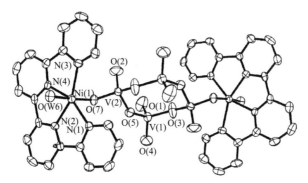

图 4-5　[{Ni(quaterpy)(H$_2$O)}$_2V_4O_{12}$] 原子连接图

Kabanos 等人在 2003 年报道了分子组成为 [V$_4$O$_8$(OCH$_3$)$_2$(μ_3-OCH$_3$)$_2$(5,5′-Me$_2$bpy)$_2$]·3CH$_3$OH 的化合物的合成和结构$^{[16]}$，该化合物也是由四个金属钒原子构成，与上述化合物不同的是，该化合物中的钒为六配位八面体构型，而且配体的 N 原子参与了钒的配位，化合物结构如图 4-6 所示。在该化合物中，4 个钒原子通过 4 个桥氧相连。另外，每个钒原子上还有一个端氧。化合物中有 4 个甲醇分子参与了配位，其中两个甲醇分子分别与一个钒原子相连，占据一个端氧的位置，另外两个甲醇分子上的氧充当三重桥氧，分别与三个钒原子配位，两个二甲基联吡啶分别与一个钒原子配位。根据配位环境的不同，钒原子可分为两类：一类连接一个端氧、两个桥氧和 CH$_3$O—基团；另一类连接一个端氧、两个桥氧、一个 CH$_3$O—基团和两个来自同一二甲基联吡啶的 N 原子。

图 4-6　$[V_4O_8(OCH_3)_2(\mu_3\text{-}OCH_3)_2(5,5'\text{-}Me_2bpy)_2]\cdot 3CH_3OH$ 的化合物结构

4.2　四钼氧簇

由于钼原子配位构型一般为六配位的要求，使得形成四钼氧簇后钼原子上的端基氧要超过两个，因而阴离子很难稳定存在。因此，尽管目前化学式为四钼氧簇的化合物报道有很多[17-30]，但仅以氧原子参与配位的具有孤立结构的四钼氧簇阴离子尚未见报道。当含氧有机配体或其他基团参与四钼氧簇阴离子的骨架形成时，由于钼原子和端基氧的键价均得到了满足，可以形成稳定的具有孤立结构的四钼氧簇阴离子。下面仅就这方面的研究工作作一介绍。

1979 年，Day 报道了组成为 $[CH_2Mo_4O_{15}H]^{3-}$ 的阴离子簇[31]，该阴离子的结构如图 4-7 所示。阴离子中的四个钼原子呈环状排列，四个钼原子位于一个平面上，每个钼原子上连接两个端基氧，钼原子之间通过一个二重桥氧相连，另外，四个钼原子还与位于钼原子所在平面下方的作为四重桥的羟基相连，同时，钼原子还分别与 $H_2CO_2^{2-}$ 基团中的两个氧原子相连。正是由于 OH^- 和 $H_2CO_2^{2-}$ 基团的参与，使得钼原子上的端基氧的个数均不超过 2，阴离子得以稳定存在。

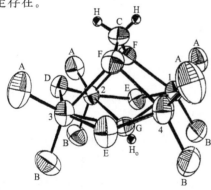

图 4-7　$[CH_2Mo_4O_{15}H]^{3-}$ 原子连接图

1985 年，Zubieta 报道了含有组成为 $[Mo_4O_8(OCH_3)_2(NNC_6H_5)_4]^{2-}$ 的阴离子的化合物[32]。在该阴离子中（图 4-8），两个四配位的钼原子和两个六配位的钼原子通过桥氧相连，四配位的钼原子上连有两个端基氧，而六配位的钼原子上除分别连有两个偶氮苯以外，还通过两个甲醇分子相互连接。同年，Zubieta 又报道了同构的 $[Mo_4O_8(OR)_2(NNC_6H_5)_4]^{2-}$

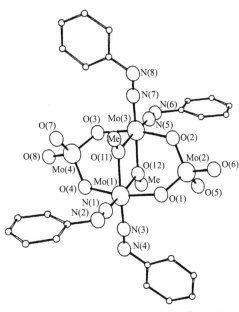

图 4-8　$[Mo_4O_8(OCH_3)_2(NNC_6H_5)_4]^{2-}$ 原子连接图

($R=CH_3$ 或 C_2H_5)[33]。其后，又报道了多个具有类似结构的化合物，在这类化合物中，核心基团是两个四配位的钼原子和两个六配位的钼原子形成的环状四钼氧簇，四个钼原子、环上的四个桥氧和两个四配位的钼原子上的端基氧保持不变，形成四钼氧簇核 $[Mo_4O_8]^{8-}$，六配位钼原子之间的桥联基团和位于端基的基团被取代，则生成新的化合物[34-40]。

1988 年，Zubieta 报道了一种新型的四钼氧簇，其阴离子组成为 $[Mo_4O_{15}(OH)(C_{14}H_8)]^{3-}$ [41]。该四钼氧簇与前述四钼氧簇结构的明显差异在于阴离子中有三个钼原子为六配位，一个钼原子为五配位（图 4-9）。

由 $(Bu_4N)_4(Mo_8O_{26})$ 和 $MeC(NH_2)NOH$ 在乙腈溶液中回流，得到分子组成为 $(Bu_4N)_2[Mo_4O_{12}\{MeC(NH_2)NO\}_2]$ 的化合物[42]，在该化合物的四钼氧簇阴离子中，其核心结构单元为 $[Mo_4O_8]^{8-}$ 基团，与前述化合物不同的是，该阴离子簇中所有的钼原子均为六配位，这是由于两个 $MeC(NH_2)NO$ 分子上的氮原子和氧原子均同时与两个钼原子配位，结构如图 4-10 所示。1994 年，Gouzerh 报道了与该阴离子同构的聚四钼氧簇阴离子 $[Mo_4O_{12}(MeCNO)_2]^{2-}$ [43]。

$[MoO_2(acac)_2]$ 和 Me_2CNOH 在室温下在甲醇溶液中反应，可得到分子组成为 $[Mo_4O_{10}(OMe)_4(Me_2CNHO)_2]$ 的化合物[43]，其分子结构如图 4-11 所示。在该化合物中，所有的钼原子均为六配位八面体构型，有两个 MeO—基团作为二重桥参与分子的组成，另两个 MeO—基团作为三重桥参与分子的组成，Me_2CNO 基团位于端基的位置。

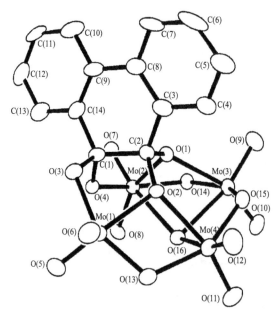

图 4-9 $[Mo_4O_{15}(OH)(C_{14}H_8)]^{3-}$ 原子连接图

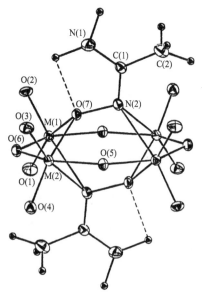

图 4-10 $[Mo_4O_{12}\{MeC(NH_2)NO\}_2]^{2-}$ 原子连接图

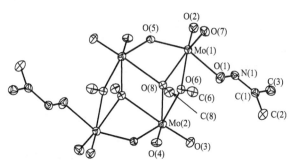

图 4-11 $[Mo_4O_{10}(OMe)_4(Me_2CNHO)_2]$ 原子连接图

Proust 报道了分子组成为 $[Bu_4N]_4[Mo_4O_{10}(NO)(OMe)(Me_2CNO)_2]$ 的化合物[44]，该化合物的聚四钼阴离子结构如图 4-12 所示。在该阴离子中，其核心结构单元为 $[Mo_4O_8]$ 基团，但由于 MeO 基团、Me_2CNO 基团和 NO 的参与配位，使得四钼簇中有 1 个钼原子为七配位五角双锥构型，2 个钼原子为六配位八面体构型，1 个钼原子为四配位四面体构型。

3-羟基吡啶酸与钼酸钠反应后，加入四丁基氯化铵，可以得到分子组成为 $[Bu_4N]_4[Mo_4O_{12}(picOH)_2]$ 的化合物[45]，其四钼簇阴离子结构如图 4-13 所示。在该阴离子中，4 个钼原子均为六配位八面体构型，4 个钼原子通过 2 个二重桥氧原子和 2 个三重桥氧原子相连，钼原子按配位环境分为两类，中间的 2 个钼原子分别连接 2 个端基氧原子、2 个二重桥氧原子和 2 个三重桥氧原子；两端的 2 个钼原子分别连接 2 个端基氧原子、1 个二重桥氧原子和 1 个三重桥氧原子，另外两个配位位置分别由吡啶环上的氮原子和羧基上的 1 个氧原子配位。

图 4-12 $[Mo_4O_{10}(NO)(OMe)(Me_2CNO)_2]^{4-}$ 原子连接图

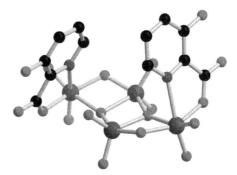

图 4-13 $[Mo_4O_{12}(picOH)_2]^{4-}$ 原子连接图

2002 年 Modec 等人报道了一系列由卤素离子和有机基团共同参与形成的聚四钼簇阴离子[46]，其原子连接方式如图 4-14 所示。由图可知，在该阴离子中，钼原子间存在相互作用。另外，每个钼原子还与六个原子或基团配位，每个钼原子连接 1 个端氧和 1 个端基卤素原子、2 个二重桥氧（或基团）和 2 个三重桥氧。

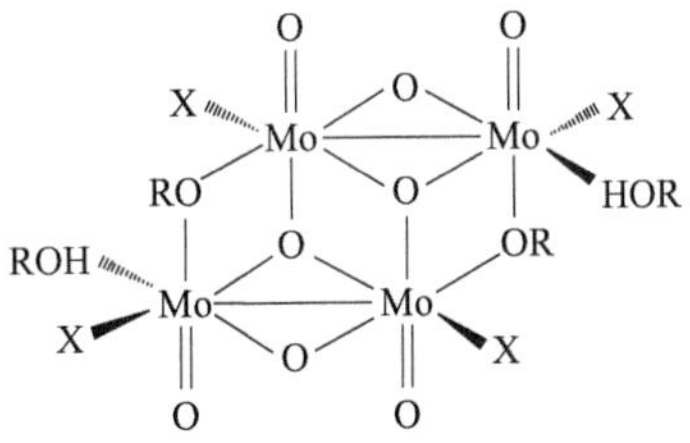

1997 年，Fink 报道了分子组成为 $[(p\text{-}Pr^iC_6H_4Me)_4Ru_4Mo_4O_{16}]$ 的化合物[47]，该化合物结构如图 4-15 所示。略去 4 个 (p-

图 4-14 $[Mo_4O_8(OR)_2(HOR)_2X_4]^{2-}$ 结构

1a R=Me, X=Cl
1b R=Et, X=Cl
1c R=Et, X=Br

$Pr^iC_6H_4Me)Ru$ 基团及钌原子间的桥氧原子，其四钼氧簇核心具有立方烷结构，每个钼原子为五配位，连接位于立方烷顶点的 3 个氧原子和 2 个端基氧原子。

2000 年，Proust 报道了分子组成为 $[\{Ru(\eta^6\text{-}p\text{-}MeC_6H_4Pr^i)\}_4Mo_4O_{16}]$ 的化合物[48]，该化合物和图 4-15 所示的化合物具有相同的核心结构。2004 年，Proust 报道了一系列由钌的芳烃配合物与钼酸钠反应制得的分子组成为 $[Ru(arene)Cl_2]_4[Mo_4O_{16}]$ [arene = $C_6H_5CH_3$、1,3,5-$C_6H_3(CH_3)_3$、1,2,4,5-$C_6H_2(CH_3)_4$] 的四钼氧簇化合物[49]。在上述系列化合物中，其核心结构单元为 $[Mo_4O_{16}]^{8-}$ 基团，其钼氧之间的连接方式如图 4-16 所示。

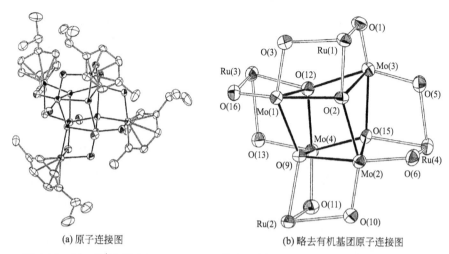

(a) 原子连接图 (b) 略去有机基团原子连接图

图 4-15　$[(p\text{-Pr}^i\text{C}_6\text{H}_4\text{Me})_4\text{Ru}_4\text{Mo}_4\text{O}_{16}]$ 原子连接图和略去有机基团原子连接图

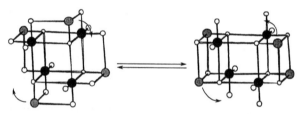

图 4-16　$[\text{Mo}_4\text{O}_{16}]^{8-}$ 中钼原子连接方式

由图可知，该四钼氧基团具有立方烷结构，4 个钼原子和 4 个氧原子分别位于立方体 8 个角的顶点上，钼原子之间通过 4 个三重桥氧相连，每个钼原子还与另外 3 个氧原子相连，其中 2 个氧原子为端基氧，另 1 个氧原子与 $[\text{Ru}(\text{arene})]^{2+}$ 基团相连，正是由于 $[\text{Ru}(\text{arene})]^{2+}$ 基团的参与配位，减少了钼原子上的端基氧个数，使得该化合物钼原子上的键价趋于合理，化合物得以稳定存在。

4.3　四钨氧簇

与四钼氧簇一样，到目前为止还没有孤立结构的四钨氧簇化合物的报道，但现有研究结果表明，在端基氧参与配位的情况下，四钨氧簇存在两种构型[50-54]，结构如图 4-17 所示。

图 4-17（a）所示的四钨氧簇阴离子具有立方烷构型，4 个钨原子位于四方体的 4 个互为对角的顶点，通过位于另外 4 个顶点的 4 个三重桥氧原子相连，每个钨原子除连接 3 个三重桥氧原子外，还连接 3 个端基氧原子；图 4-

17(b) 所示的聚四钨氧簇阴离子具有 C_{2h} 对称性，4 个钨原子分为两类，钨原子之间通过 4 个桥氧原子相连成环，然后通过 2 个三重桥氧原子相连。其中 2 个钨原子分别连接 2 个二重桥氧、2 个三重桥氧和 2 个端基氧原子；另 2 个钨原子分别连接 2 个二重桥氧、1 个三重桥氧和 3 个端基氧原子。无论是具有 T_d 对称性还是具有 C_{2h} 对称性的四钨氧簇阴离子都非常少见，已有的几例也需有四钨氧簇阴离子上的端基氧原子参与配位，这是由于在这两个阴离子簇中都包含有非常独特的结构现象，即阴离子中钨原子连有三个端基氧原子。

向 20mL 含有 0.82g 钨酸钾的水溶液中加入 10mL 含有 0.35g 碳酸钾的水溶液，然后加入 2.5mL 30%的 H_2O_2，可得到分子组成为 $K_6[W_4O_8(O_2)_6(CO_3)] \cdot 6H_2O$ 的化合物[55]。该化合物的四钨氧簇阴离子具有如图 4-18 所示的结构。由图可知，该阴离子中的 4 个钨原子通过 2 个三重桥氧、2 个二重桥氧和 1 个碳酸根桥相连，钨原子上端基氧的位置部分由 O_2^{2-} 基团取代。

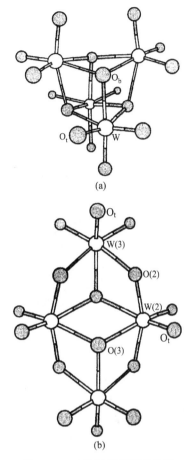

图 4-17　四钨氧簇的两种构型

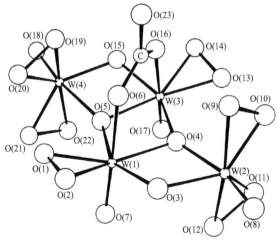

图 4-18　$[W_4O_8(O_2)_6(CO_3)]^{6-}$ 结构

将1.25g水合三氧化钨加入4mL 30%的H_2O_2中，50℃下搅拌至溶液无色，离心分离掉不溶物，得到的溶液用2mol/L H_2SO_4调pH值到1.5，滴加2.78g氯化四丁基铵溶于10mL水中的溶液，然后加入5mL乙醇并冷却至4℃，可得到无色晶体。元素及晶体结构分析表明，化合物分子式为$[Bu_4N]_2[W_4O_6(O_2)_6(OH)_2(H_2O)_2]$[56]，其阴离子结构如图4-19所示。4个钨原子通过2个二重桥氧和2个三重桥氧相连，钨原子上的端氧部分由H_2O和O_2^{2-}基团取代。

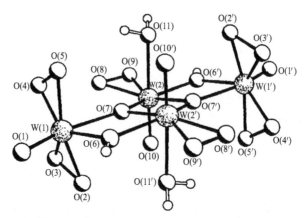

图4-19 $[W_4O_6(O_2)_6(OH)_2(H_2O)_2]^{2-}$结构

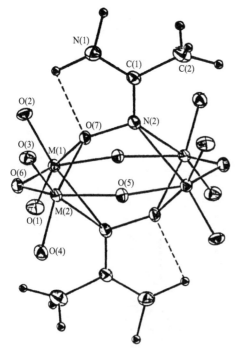

图4-20 $[W_4O_{12}\{MeC(NH_2)NO\}_2]^{2-}$结构

将$[Bu_4N]_2[W_6O_{19}]$ 3.788g加到含有1.776g $MeC(NH_2)NOH$的25mL甲醇溶液中，回流20h，得到分子组成为$[Bu_4N]_2[W_4O_{12}\{MeC(NH_2)NO\}_2]\cdot 2MeOH$的化合物[42]，结构分析表明，该化合物中四钨氧簇阴离子结构如图4-20所示。4个钨原子通过4个二重桥氧相连构成1个四元环，每个钨原子上连接2个端基氧，2个$MeC(NH_2)NO$基团中的氮和氧分别从八元环的上方和下方与钨原子配位，钨原子均为六配位八面体构型。

1997年Finke等人报道了组成为$2CH_3CN\subset[Bu_4N]_2[\{Ir(1,5-COD)\}_6W_4O_{16}]\cdot 2CH_3CN$的包合物[57]。该包合物阴离子含四钨氧簇核$[W_4O_{16}]^{8-}$，6个$[Ir(1,5-COD)]^+$有机金属基团与两个端基氧

相连，四个钨原子呈四面体排列，6个铱原子位于 $[W_4O_4]$ 基团构成的立方体面的上方（图 4-21）。$[W_4O_{16}]^{8-}$ 单元中的每个钨原子有 3 个端基氧原子，这违背了 Lipscomb 规则，但是 6 个 $[Ir(1,5\text{-COD})]^+$ 基团通过 2 个端基氧原子分别连接到 2 个钨原子上，3 个端基氧形成的 Ir—O 键使得聚四钨氧簇阴离子得以稳定存在。

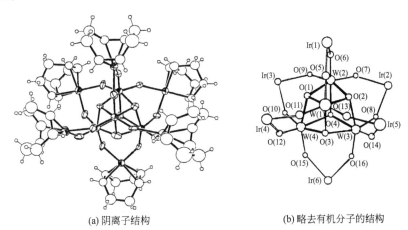

(a) 阴离子结构　　　　　　　(b) 略去有机分子的结构

图 4-21　$[\{Ir(1,5\text{-COD})\}_6W_4O_{16}]^{2-}$ 结构和略去有机分子的结构

参考文献

[1] Fuchs J, Mahjour S, Pickardily J. Structure of the "true" metavanadate ion. Angew Chem Int Ed Engl, 1976, 15 (6)：374-375.

[2] Millet J M, Parker H S, Roth R S. Syntheses and unit-cell determination of $Ba_3V_4O_{13}$ and low-and high-temperature $Ba_3P_4O_{13}$. J Amer Ceram Soc, 1986, 69：C103-C105.

[3] Gatehouse B M, Guddat L W, Roth R S. The crystal structure of $Ba_3V_4O_{13}$. J Solid State Chem, 1987, 71：390-395.

[4] Yang G Y, Gao D W, Chen Y, et al. $[Ni(C_{10}H_8N_2)_3]_2[V_4O_{12}] \cdot 11H_2O$. Acta Cryst, 1998, C54：616-618.

[5] Nakano H, Ozeki T, Yagasaki A. $(Et_4N)_4[V_4O_{12}] \cdot 2H_2O$. Acta Cryst, 2002, C58：m464-m465.

[6] Kucsera R, Gyepes R, Zurkova L. Synthesis and crystal strucure of $[Ni(phen)_3]_2$ $[V_4O_{12}] \cdot 7H_2O$. Solid State Phenomena, 2003, 90-91：329-334.

[7] Qi Y J, Wang Y H, Li H M, et al. Hydrothermal syntheses and crystal structures of bi-metallic cluster complexes $[\{Cd(phen)_2\}_2V_4O_{12}] \cdot 5H_2O$ and $[Ni(phen)_3]_2 \cdot 17.5H_2O$. J Mol Struct, 2003, 650：123-129.

[8] Huang M H, Bi L H, Dong S J. Bis [tris(2,2'-bipyridyl-$\kappa^2 N,N'$)iron(Ⅱ)] cyclo-tetra-

[9] Zhang Y P, Zapf P J, Meyer L M, et al. Polyoxoanion coordination chemistry: synthesis and characterization of the heterometallic, hexanuclear clusters [{Zn(bipy)$_2$}$_2$V$_4$O$_{12}$], [{Zn(phen)$_2$}$_2$V$_4$O$_{12}$]·H$_2$O, and [{Ni(bipy)$_2$}$_2$Mo$_4$O$_{14}$]. Inorg Chem, 1997, 36: 2159-2165.

[10] Day V W, Klemperer W G, Yagasaki A. Synthesis and structure of the new organometallic polyoxovanadates. Chem Lett, 1990, 19 (8): 1267-1270.

[11] Hayashi Y, Miyakoshi N, Shinguchi T, et al. A stepwise growth of polyoxovanadate by reductive coupling reaction with organometallic palladium complex: formation of [{(η^3-C$_4$H$_7$)Pd}$_2$V$_4$O$_{12}$]$^{2-}$, [V$_{10}$O$_{26}$]$^{4-}$ and [V$_5$O$_{36}$(Cl)]$^{4-}$. Chem Lett, 2001: 170-171.

[12] Lu Y, Wang E B, Yuan M, et al. Hydrothermal synthesis and X-ray single crystal structure of bimetallic cluster complex [{Co(phen)$_2$}$_2$V$_4$O$_{12}$]·H$_2$O. J Mol Struct, 2002, 607: 189-194.

[13] Kucsera R, Gyepes R, Zurkova L. The crystal structure of the cluster complex [{Co(phen)$_2$}$_2$V$_4$O$_{12}$]·H$_2$O. Cryst Res Technol, 2002, 37 (8): 890-895.

[14] Sun X Z, Wang B Y, Zhong L Y. μ-Tetravanadato-bis[bis(2,2'-bipyridine)-cadmium(II)]. Acta Cryst, 2004, E60: m1100-m1102.

[15] Xiao D R, Wang E B, Lü J, et al. Dehydrogenative coupling of 2,2'-bipyridine: hydrothermal synthesis and crystal structure of a novel polyoxovanadate decorated with the 2,2'; 6',2"; 6",2"'-quaterpyridine ligand. Inorg Chem Commun, 2004, 7: 437-439.

[16] Manos M J, Tasiopoulos A J, Tolis E J, et al. A new class of ferromagnetically-coupled mixed valence vanadium (IV/V) polyoxometalaes. Chem Eur J, 2003, 9 (3): 695-703.

[17] Jahr P K F, Fuchs D J. New methods and results in the study of polyacids. Angew Chem Inter Edit, 1966, 5 (8): 689.

[18] Gatehouse B M, Leverett P. The chain structure of potassium tetramolybdate, K$_2$Mo$_4$O$_{13}$. Chem Commun, 1970, 740.

[19] Gatehouse B M, Miskin B K. Structural studies in the Li$_2$MoO$_4$-MoO$_3$ system: part 1. The low temperature form of lithium tetramolybdate, L-Li$_2$Mo$_4$O$_{13}$. J Solid State Chem, 1974, 9: 247.

[20] Gatehouse B M, Miskin B K. Crystal structure of potassium tetramolybdate, K$_2$Mo$_4$O$_{13}$, and its relationship to the structure of other univalent metal polymolybdate. J Chem Soc (A), 1971 (13): 2107-2112.

[21] Tytko K H, Schönfeld B. Über isopolymolybdatfeststoffe und deren beziehung zu isopolymolybdationen in wäbriger lösung. Z Naturforsch, 1975, 30b: 471-484.

[22] Gatehouse B M, Miskin B K. Structural studies in the Li$_2$MoO$_4$-MoO$_3$ system: Part 2. The low temperature form of lithium tetramolybdate, L-Li$_2$Mo$_4$O$_{13}$. J Solid State Chem,

1975, 15: 274.

[23] Tytko K H. Über den mechanismus der bildung von polyanionen in wäbriger lösung. Zum bildungsmechanismus eines Polytetramolybdations $[Mo_4O_{14}^{4-}]_n$. Z Naturforsch, 1976, 31b: 737-748.

[24] Goiffon A, Spinner B. Édifications structurales des isopolyanions du molybdène Ⅵ de condensation croissante et décroissante au cours de l'acidification. Nouvelle interprétation. Bulletin De La Sociéyé Chimque De France, 1977, 11-12: 1081.

[25] Toléd P P, Touboul M. Structure cristalline du tetramolybdate de Thallium (Ⅰ), $Tl_2Mo_4O_{13}$. Acta Cryst, 1978, B34: 3547.

[26] D'Yachenko O G, Tabacenko V V, Sundberg M. Crystal structure of $(Mo, W)_9O_{25}$, homologue of the Mo_4O_{11} (orthorhombic)-type structure. J Solid State Chem, 1995, 119: 8.

[27] Marrot J, Savariault J M. Two original infinite chain in the new caesium tetramolybdate compound $Cs_2Mo_4O_{13}$. Acta Cryst, 1995, C21: 2201.

[28] Guillou N, Férey G, Whittingham M S. $Mo^V - Mo^{VI}$ cationic ordering in the layered molybdate $(C_2H_{10}N_2)[Mo_4O_{12}]$. J Mater Chem, 1998, 8 (10): 2277.

[29] Harrison W T A. Strontium tetramolybdate dehydrate $SrMo_4O_{13} \cdot 2H_2O$. Acta Cryst, 1999, C5: 485.

[30] Randy S Rarig Jr, Pamela J Hagrman, Jon Zubieta. Ligand influences on the structures of copper molybdate chains: hydrothermal synthesis and structural characterizations of $[Cu(2,2'-bipyridine)Mo_4O_{13}]$ and $[Cu(2,3-bis(2-pyridyl)pyrazine)Mo_2O_7]$. Solid State Sci, 2002, 4 (1): 77-82.

[31] Day V W, Fredrich M F, Xlemperer W G, Liu R S. Polyoxomolybdate-hydrocarbon interactions. Synthesis and structure of the $[CH_2Mo_4O_{15}H]^{3-}$ anion and related methylenedioxymolybdates. J Am Chem Soc, 1979, 101 (2): 491-492.

[32] Hsieh T C, Zubieta J. Synthesis and characterization of a tetranuclear oxomolybdate containing coordinatively bound diazenido units $[Mo_4O_8(OCH_3)_2(NNC_6H_5)_4]^{2-}$: a versatile precursor for the synthesis of complexes with the $[Mo(NNC_6H_5)_2]^{2+}$ unit. Inorg Chem, 1985, 24: 1287-1288.

[33] Hsieh T C, Zubieta J A. Synthesis and chemical characterization of diazenido derivatives of oxomolybdate clusters: the structures of the tetranuclear complexes $[Mo_4O_8(OR)_2(NNC_6H_5)_4]^{2-}$ ($R = CH_3$ or C_2H_5). Polyhedron, 1986, 5 (1-2): 305-314.

[34] Shalkh S N, Zubieta J. New organonitrogen-deribatized polyoxomolybdate anion clusters. Syntheses and characterization of tetranuclear and octanuclear oxomolybdates containing coordinatively bound organohydrazido (2-) units: $[Mo_4O_{10}(OCH_3)_2(NNPh_2)_2]^{2-}$ and $[Mo_8O_{16}(OCH_3)_6(NNMePh)_6]^{2-}$. Inorg Chem, 1986, 25 (26): 4613-4615.

[35] Burkholder E, Zubieta J. Hydrothermal synthesis and structure of [Ni(tpyrpyz)$_2$]$_2$[Mo$_4$O$_{12}$F$_2$][Mo$_6$O$_{19}$]·2H$_2$O, a material exhibiting an unusual tetranuclear polyoxofluoromolybdate cluster [Mo$_4$O$_{12}$F$_2$]$^{2-}$. Inorg Chim Acta, 2004, 357: 279-284.

[36] Hsieh T C, Shaikh S N, Zubieta J. Derivatized polyoxomolybdates. Synthesis and characterization of oxomolybdate clusters containing coordinatively bound diazenido units. Crystal and molecular structure of the octanuclear oxomolybdate (NHEt$_3$)$_2$(n-Bu$_4$N)$_2$[Mo$_8$O$_{20}$(NNPh)$_6$] and comparison to the structures of the parent oxomolybdate α-(n-Bu$_4$N)$_4$[Mo$_8$O$_{26}$] and the tetranuclear (diazenido) oxomolybdates (n-Bu$_4$N)$_2$[Mo$_4$O$_{10}$(OMe)$_2$(NNPh)$_2$] and (n-Bu$_4$N)$_2$[Mo$_4$O$_8$(OMe)$_2$(NNC$_6$H$_4$NO$_2$)$_4$]. Inorg Chem, 1987, 26: 4079-4089.

[37] Liu S C, Shaikh S N, Zubieta J. Coordination complexes of polyoxomolybdate anions. Characterization of a tetranuclear core from reactions in methanol: syntheses and structures of two polyoxomolybdate alcoholates, (MePPh$_3$)$_2$[Mo$_4$O$_{10}$(OCH$_3$)$_6$] and (n-Bu$_4$N)$_2$[Mo$_4$O$_{10}$(OCH$_3$)$_2$(OC$_6$H$_4$O)$_2$], and their relationship to a general class of tetranuclear cluster types [Mo$_4$O$_x$(OMe)$_2$(L)$_y$(LL)$_z$]$^{2-}$. Inorg Chem, 1987, 26: 4303-4305.

[38] Shaikh N, Zubieta J. Synthesis and structural characterization of a tetramolybdate, [Mo$_4$O$_{12}$(C$_8$H$_6$N$_4$)]$^{2-}$, containing a bridging phthalazine-1-hydrazido (2-) ligand. Comparison to the structure of [Mo$_4$O$_{10}$(OMe)$_2$(NNMePh)$_2$]$^{2-}$, a tetramolybdate species exhibiting ligation to a terminal monodentate hydrazido (2-) group. Inorg Chem, 1988, 27: 1896-1903.

[39] Attanasio D, Fares V, Imperatori P. Synthesis and structure of a mixed-valence tetramolybdate containing a binucleating diazene legand. J Chem Soc Chem Commun, 1986: 1476-1477.

[40] Roh S-G, Proust A, Robert F, Gouzerh P. Coordination chemistry of amidoximes. Molybdenum complexes of (alkyl) aminoacetamidoximes and fumaromonoamidoxime. Inorg Chim Acta, 2003, 342: 311-315.

[41] Liu S C, Shaikh S N, Zubieta J. Polyoxomolybdate-o-benzoquinone interactions. Synthesis and structure of a diacetal derivative, [Mo$_4$O$_{15}$(OH)(C$_{14}$H$_8$)]$^{3-}$, from 9,10-phenanthrenequinone carbonyl insertion. Comparison to the reaction products with tetrachloro-1,2-benzoquinone, the ligand-bridged binuclear complexes [(MoO$_2$Cl$_2$)$_2$L]$^{2-}$, L=(C$_6$Cl$_2$O$_4$)$^{2-}$ and (C$_2$O$_4$)$^{2-}$, formed via carbonyl insertion and chloride transfer. Inorg Chem, 1988, 27: 3064-3066.

[42] Chilou V, Gouzerh P, Jeannin Y, et al. [Mo$_4$O$_{12}$ {MeC(NH$_2$)NO}$_2$]$^{2-}$ (M=Mo or W), a tetranuclear complex with a μ_4-acetamidoximate ligand as an unprecedented bridge. J Chem Soc Chem Commun, 1987: 1469-1470.

[43] Proust A, Gouzerh P, Robert F. Reactivity of acetone oxime towards oxomolybdenum

[43]　(Ⅵ) complexes. Part 1. Syntheses and crystal structures of tetranuclear molybdenum (Ⅵ) complexes. J Chem Soc Dalton Trans, 1994: 819-824.

[44]　Proust A, Gouzerh P, Robert F. Reactivity of acetone oxime towards oxomolybdenum (Ⅵ) complexes. Part 2. Syntheses, crystal structures and reactivity of molybdenum nitrosyl complexes. J Chem Soc Dalton Trans, 1994: 825-833.

[45]　Quintal S M O, Nogueira H I S, Carapuca H M, et al. Polynuclear molybdenum and tungsten complexes of 3-hydroxy-picolinic acid and the crystal structures of $(^nBu_4N)_2$ $[Mo_4O_{12}(picOH)_2]$ and $(^nHex_4N)_2[Mo_2O_6(picOH)_2]$. J Chem Soc Dalton Trans, 2001: 3196-3201.

[46]　Modec B, Brenčič J V, Zubieta J. A templated synthesis of tetranuclear polyoxoalkoxymolybdates (Ⅴ). Bromo coordinated oxomolybdenum (Ⅴ) clusters: known core structure with new ligands. Oxidation to the lindquist ainon. J Chem Soc Dalton Trans, 2002: 1500-1507.

[47]　Fink G S, Plasseraud L, Ferrand V, et al. $[(p-Pr^iC_6H_4Me)_4Ru_4Mo_4O_{16}]$: an amphiphilic organoruthenium oxomolybdenum cluster presenting a unique framework geometry. Chem Commun, 1997: 1657-1658.

[48]　Artero V, Proust A, Herson P, et al. (η^6-Arene) ruthenium oxomolybdenum and oxotungsten clusters. Stereochemical non-rigidity of $[\{Ru(\eta^6-p-MeC_6H_4Pr^i)\}_4Mo_4O_{16}]$ and crystal structure of $[\{Ru(\eta^6-p-MeC_6H_4Pr^i)\}_4W_2O_{10}$. Chem Commun, 2000: 883-884.

[49]　Laurencin D, Fidalgo E G, Villanneau R, et al. Framework fluxionality of organometallic oxides: synthesis, crystal structure, EXAFS, and DFT studies on $[\{Ru(\eta^6-arene)\}_4Mo_4O_{16}]$ complexes. Chem Eur J, 2004, 10: 208-217.

[50]　Skarstad P M, Geller S. $(W_4O_{16})^{8-}$ polyion in the high temperature modification of silver tungstate. Mat Res Bull, 1975, 10: 791-800.

[51]　Chan L Y Y, Geller S. Crystal structure and conductivity of 26-silver 18-iodide tetratungstate, $Ag_{26}I_{18}W_4O_{16}$. J Solid State Chem, 1977, 21: 331-347.

[52]　Bridgeman A J, Cavigliasso G. Molecular and electronic structures of six-coordinate W complexes and polyanions containing tri-oxo groups. Polyhedron, 2001, 20: 3101-3111.

[53]　Eva M Voigt. Charge-transfer spectra in nonpolar solvents. J Phys Chem, 1966, 70: 598-600.

[54]　Bridgeman A J, Cavigliasso G. Bonding in $[W_4O_{16}]^{8-}$ isopolyanions. Polyhedron, 2002, 21: 2201-2206.

[55]　Stomberg R. The crystal structure of potassium μ-carbonato-octaoxohexaperoxo tetratungstate (Ⅵ)-6-water, $K_6[W_4O_8(O_2)_6(CO_3)] \cdot 6H_2O$, a new type of polytungstate. Acta Chem Scand, 1985, A39: 507-514.

[56] Griffith W P, Parkin B C, White A J P, et al. The crystal structures of $[NMe_4]_2[(Ph-PO_3)\{MoO(O_2)_2\}_2\{MoO(O_2)_2(H_2O)\}]$ and $[NBu_4^n]_2[W_4O_6(O_2)_6(OH)_2(H_2O)_2]$ and their use as catalytic oxidants. J Chem Soc Dalton Trans, 1995: 3131-3138.

[57] Hayashi Y, Müller F, Lin Y, et al. $2CH_3CN(n\text{-}Bu_4N)_2[\{Ir(1,5\text{-}COD)\}_6W_4O_{16}]\cdot 2CH_3CN$: a hybrid inorganic-organometallic, flexible cavity host, acetonitrile-guest complex composed of a $[W_4O_4]^{n+}$ tetrarungstate cube and six polyoxoanion-supported (1,5-COD) Ir^+ organometallic groups. J Am Chem Soc, 1997, 119: 11401-11407.

第5章 五金属氧簇

五金属氧簇同样存在着分子式为五金属氧簇,而实际结构为多维化合物的情况[1-10]。但到目前为止,具有孤立结构的五钒氧簇、五钼氧簇和五钨氧簇均有报道。

5.1 五钒氧簇

分子组成含有五钒氧簇阴离子 $[V_5O_{14}]^{3-}$ 的化合物在 20 世纪 70 年代就已报道[11,12],但其结构在 1989 年才由 Klemperer 和 Day 确定[13]。

5.1.1 合成

搅拌下向 4.5mL 0.41mol/L 的 $[(n\text{-}C_4H_9)_4N]OH$ 乙腈溶液中加入 0.97g $[(n\text{-}C_4H_9)_4N]_3H_3V_{10}O_{28}$ 溶于 25mL 乙腈的溶液,由 ^{51}V 核磁谱可知,得到的溶液中至少含有四种不同的钒氧簇离子,溶液呈深橙色,过滤掉不溶物,然后将溶液加热至沸腾,在 10~15min 内将溶液蒸发至 15~20mL,溶液颜色由深橙转变为无色。

$$H_3V_{10}O_{28}^{3-} + 3OH^- \xrightarrow{\triangle} 2V_5O_{14}^{3-} + 3H_2O$$

室温搅拌下，向反应液中加入 30～40mL 乙醚，得到白色沉淀，将白色沉淀溶于 15mL 丙酮，然后再加入 2～3mL 乙酸乙酯，在 -7℃ 放置 12h，可以得到晶体。

5.1.2 谱学表征[13]

化合物 $[(C_4H_9)_4N]_3V_5O_{14}$ 的红外光谱在 450～1000cm^{-1} 范围内有如下振动吸收峰：474cm^{-1}（w）、530cm^{-1}（w）、814cm^{-1}（w）、886cm^{-1}（s）、841cm^{-1}（s）、967cm^{-1}（s）。25℃，0.02mol/L 的 $[(C_4H_9)_4N]_3V_5O_{14}$ 丙酮溶液，^{51}V 核磁谱相对外标 $VOCl_3$ 的化学位移为 -539cm^{-1}（3V）和 -613cm^{-1}（2V）。

5.1.3 晶体结构[13]

五钒氧簇阴离子结构如图 5-1 所示，该阴离子由五个钒氧四面体通过共顶点相连而成。整个阴离子簇可看作三角双锥结构，钒原子位于三角双锥的 5 个顶点，2 个位于轴向位置的钒原子连接 3 个桥氧和 1 个端基氧原子，3 个位于纬向的钒原子分别连接 2 个桥氧和 2 个端基氧原子。整个阴离子具有 D_{3h} 对称性。3 个连接两个端基氧的钒原子及它们连接的 6 个端基氧原子位于同一平面内，平均偏差 0.07Å。端基氧原子钒氧 V—O_t 键长为 1.58～1.60Å，桥氧原子钒氧 V—O_b 键长在 1.723～1.818Å，O—V—O 键角为 107.6°～112.7°。

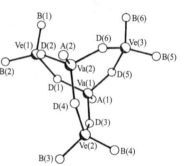

图 5-1 五钒氧簇 $[V_5O_{14}]^{3-}$ 原子连接图

5.2 五钼氧簇

5.2.1 合成

1977 年 Klemperer 报道了分子组成为 $[(n\text{-}C_4H_9)_4N]_3[Mo_5O_{17}H]$ 的五钼氧簇化合物的合成[14]。将 1g $[(n\text{-}C_4H_9)_4N]_2Mo_2O_7$ 溶于 1mL 水和 10mL 乙腈的混合溶剂中，得到澄清溶液，快速搅拌下向该溶液中缓慢（15mL/min）加入 80mL 乙醚，得到分子组成为 $[(n\text{-}C_4H_9)_4N]_3[Mo_5O_{17}H]$ 的无定形沉淀。上述反应可由下式描述：

$$7Mo_2O_7^{2-} + H_2O \rightleftharpoons 2[Mo_5O_{17}H]^{3-} + 4MoO_4^{2-}$$

$[(n\text{-}C_4H_9)_4N]_3Mo_5O_{17}H$ 在溶液中不稳定,在乙腈溶液中存在下列平衡:

$$2[Mo_5O_{17}H]^{3-} \rightleftharpoons \alpha\text{-}Mo_8O_{26}^{4-} + Mo_2O_7^{2-} + H_2O$$

Klemperer 对其在溶液中的不稳定性给出了解释,认为 $[Mo_5O_{17}H]^{3-}$ 在溶液中的不稳定性源自其结构 $(MoO_4^{2-})(OH^-)(Mo_4O_{12})$,五钼氧簇中 OH^- 基团的存在使其有可能在溶液中解离出质子:

$$2(MoO_4^{2-})(OH^-)(Mo_4O_{12}) \longrightarrow 2MoO_4^{2-} + 2H^+ + 2(O^{2-})(Mo_4O_{12})$$

解离出的质子与溶液中的碱性钼酸根基团发生缩合反应:

$$2MoO_4^{2-} + 2H^+ \longrightarrow Mo_2O_7^{2-} + H_2O$$

$(O^{2-})(Mo_4O_{12})$ 基团二聚,进而发生异构化:

$$2(O^{2-})(Mo_4O_{12}) \longrightarrow [(O^{2-})(Mo_4O_{12})]_2$$

$$[(O^{2-})(Mo_4O_{12})]_2 \longrightarrow (MoO_4^{2-})_2(Mo_6O_{18})$$

二聚阴离子 $[(O^{2-})(Mo_4O_{12})]_2$ 即 $\beta\text{-}Mo_8O_{26}^{4-}$。最后 $\beta\text{-}Mo_8O_{26}^{4-}$ 异构化为 $\alpha\text{-}Mo_8O_{26}^{4-}$。

5.2.2 结构

由于化合物在溶液中不稳定,因此难以得到化合物的晶体结构,溶液方法也不能得到化合物的结构信息。Klemperer 等人通过元素分析和红外数据,提出了该化合物中五钼氧簇阴离子的可能结构,如图 5-2 所示。阴离子具有 C_{2V} 对称性,4 个钼原子通过 4 个桥氧相连构成一个四元环,每个钼原子上另外连接 2 个端基氧,第五个钼原子通过 2 个三重桥氧在四元环平面的一侧与四元环上的四个钼原子相连,另外连接两个端基氧;在四元环平面的另一侧,羟基与四个钼原子相连,四元环上的钼原子为六配位八面体构型,第五个钼原子为四配位四面体构型。

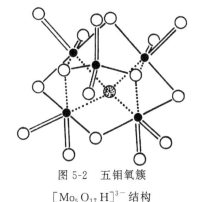

图 5-2 五钼氧簇 $[Mo_5O_{17}H]^{3-}$ 结构

5.2.3 相关化合物

2004 年我们报道了一个由 4,4′-联吡啶桥联二聚的五钼氧簇[15],分子组成为 $[4,4'\text{-}H_2bpy]_3[4,4'\text{-}Hbpy]_2[\{Mo_5O_{16}(OH)\}_2(4,4'\text{-}bpy)]$。该五钼氧簇的结构与 Klemperer 提出的聚五钼氧簇阴离子的结构颇为相似,其不对称结构单元如图 5-3 所示。四个钼原子与 4 个二重桥氧相连形成四元环,每一个钼原子上另外连接两个端基氧，第五个钼原子通过 2 个三重桥氧与环上的 4 个钼原子相连，同时还连接两个端基氧原子，环的另一侧一个作为四重桥的羟基与 4 个钼原子相连，与 Klemperer 报道的化合物结构的区别在于，该化合物的第五个钼原子与 4,4′-联吡啶的氮配位，具有三角双锥构型。

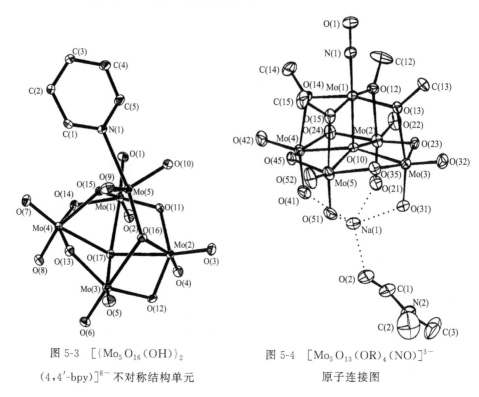

图 5-3 $[\{Mo_5O_{16}(OH)\}_2$
$(4,4'\text{-bpy})]^{8-}$ 不对称结构单元

图 5-4 $[Mo_5O_{13}(OR)_4(NO)]^{3-}$
原子连接图

1993 年，Gouzerh 等人报道了阴离子为五钼氧簇核的化合物[16]，其阴离子组成为 $[Mo_5O_{13}(OR)_4(NO)]^{3-}$（R=CH$_3$，C$_2H_5$），阴离子结构如图 5-4 所示。该阴离子可以看作单缺位的 Lindqvist 结构阴离子。其中的一个 MoO$_6$ 基团被 Mo(NO)$^{3+}$ 基团取代，甲氧基取代了连接 Mo(NO)$^{3+}$ 基团和其他钼原子的桥氧，缺位位置由 Na$^+$ 占据，四个 Mo(Ⅵ) 中心共平面，最大偏差 0.0012Å，中心氧原子在 4 个 Mo(Ⅵ) 构成的平面上，向 Mo(NO)$^{3+}$ 基团偏移 0.4Å，整个阴离子具有 C_{4V} 对称性。Gouzerh 小组在后续的研究工作中，又基于 $[Mo_5O_{13}(Me)_4(NO)]^{3-}$，将过渡金属离子和金属有机基团与该阴离子相结合，得到了多个具有新颖结构的化合物[17-19]。图 5-5～图 5-12 列出了该系列化合物的结构图。由相关化合物的研究可以发现，尽管五钼氧簇本身的稳定性很弱，但以五钼氧簇构成的化合物，特别是五钼氧簇担载金属有机基团形成的化合物却展现出结构的多样性。

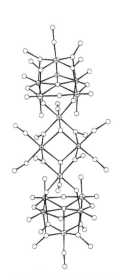

图 5-5 [{Ni(MeOH)$_2$}$_2${Mo(NO)}$_2$(μ_3-OH)$_2$(μ-OMe)$_4${Mo$_5$O$_{13}$(OMe)$_4$(NO)}$_2$]$^{2-}$ 结构

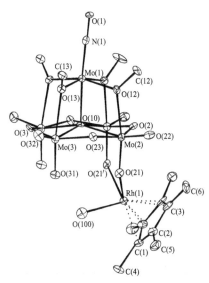

图 5-6 [{Cp*Rh(H$_2$O)}Mo$_5$O$_{13}$(OMe)$_4$(NO)]$^-$ 原子连接图

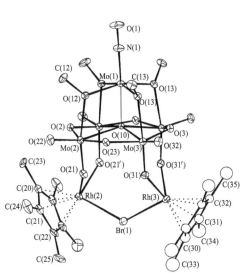

图 5-7 [{(Cp*Rh)$_2$(μ-Br)}Mo$_5$O$_{13}$(OMe)$_4$(NO)] 原子连接图

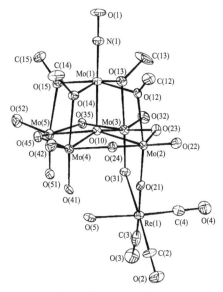

图 5-8 [Re(CO)$_3$(H$_2$O){Mo$_5$O$_{13}$(OMe)$_4$(NO)}]$^{2-}$ 原子连接图

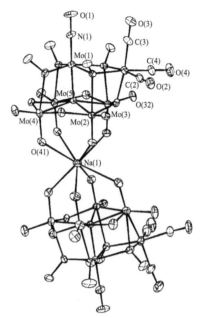

图 5-9 $[Na\{Mo_5O_{13}(OMe)_4(NO)\}_2\{Mn(CO)_3\}_2]^{3-}$ 原子连接图

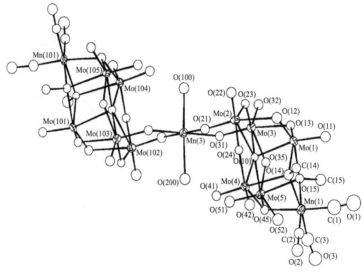

图 5-10 $[Mn(H_2O)_2\{Mo_5O_{16}(OMe)_2\}_2\{Mn(CO)_3\}_2]^{4-}$ 原子连接图

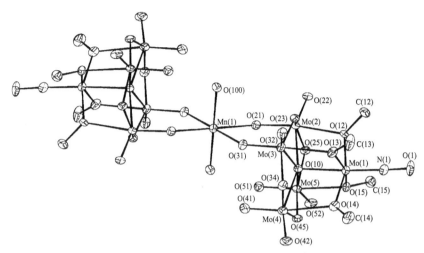

图 5-11 $[Mn(H_2O)_2\{Mo_5O_{13}(OMe)_4(NO)\}_2]^{4-}$ 原子连接图

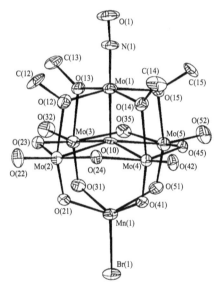

图 5-12 $[Re(CO)_3(H_2O)\{Mo_5O_{13}(OMe)_4(NO)\}]^{2-}$ 原子连接图

5.3 五钨氧簇

早在 1904 年 Schaefer 就报道 1 份的氯化铷和 3~3.5 份的钨酸反应可制得五钨酸铷 $Rb_2W_5O_{16}$[20]，但限于当时的条件，他没有给出化合物的结构。1913 年 Pereira 提出将乙酸加到 Na_2WO_4 溶液中，很可能形成五钨酸根[21]。1968 年，Marie 用传统的方法以六氢吡啶和钨化合物为原料反应，得到了分子组成为

$(C_5H_{12}N)_2W_5O_{15}$ 的化合物[22]。但这些报道均未确定化合物的结构。

5.3.1 合成

1996 年，Hartl 报道了一种结构明确的新型五钨氧簇，其分子组成为 $K_7HW_5O_{19} \cdot 10H_2O$[23]，化合物由 11.2g（0.0343mol）K_2WO_4 和 1.25g（0.005mol）$WO_3 \cdot H_2O$ 在 12.5mL 20℃除去 CO_2 的水溶液中封闭搅拌，溶液的酸化度为 $H^+/WO_4^{2+}=0.2$，溶液中钨酸根浓度大约为 2mol/L，体系从最初的黄色悬浮液变为几乎白色，然后过滤，从滤液中可得到在空气中稳定的单晶。

5.3.2 结构

化合物 $K_7HW_5O_{19} \cdot H_2O$ 单晶结构分析表明，阴离子由 5 个 WO_6 八面体构成，其中两个八面体含有游离平面，即每个平面含有 3 个相邻的端基氧原子，2 个八面体具有一个包含 2 个端基氧原子的游离棱，1 个八面体只有 1 个端基氧原子（图 5-13）。

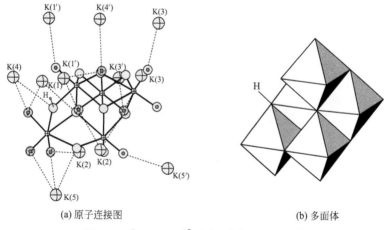

(a) 原子连接图　　　　　　(b) 多面体

图 5-13　$[HW_5O_{19}]^{7-}$ 原子连接图和多面体

五钨氧簇阴离子 $[HW_5O_{19}]^{7-}$ 中存在 2 个含有 3 个端基氧原子的 WO_6 八面体，不符合 Lipscomb 规则，正常情况下难以稳定存在，但在该化合物中由于晶体中钾离子的配位和与结晶水分子形成的氢键起到了稳定化作用，使其在晶体中稳定；在溶液中，这种端基氧基团 $fac\text{-}[MO_3]$（M=Mo，W）非常不稳定且易于反应，端基氧原子的数量由于与其他原子结合或构型转变而减少。

5.3.3 相关化合物

Lindqvist 结构的六钨氧簇阴离子 $[W_6O_{19}]^{2-}$ 失去 1 个 WO 基团可以形成

缺位的五钨氧簇 $[W_5O_{18}]^{6-}$，该阴离子具有很强的反应活性，可以作为四齿配体与金属离子或基团反应形成新型化合物。

1971 年 Weakley 首次报道了由 $[W_5O_{18}]^{6-}$ 和稀土离子形成的一系列化合物 $[LnW_{10}O_{35}]^{7-}$（Ln＝La，Ce，Pr，Nd，Sm，Ho，Er，Yb 和 Y）[24,25]，并对化合物进行了分析，后来证明阴离子的准确组成应该是 $[LnW_{10}O_{36}]^{9-}$。1974 年，Weakley 确定了化合物 $Na_6CeW_{10}O_{36} \cdot 30H_2O$ 的晶体结构[26]。在该化合物中，$[W_5O_{18}]^{6-}$ 基团作为四齿配体与 Ce^{3+} 配位，五个 $[W_5O_{18}]^{6-}$ 基团和一个 Ce^{3+} 构成具有夹心结构的 $[CeW_{10}O_{36}]^{9-}$（图 5-14）。Yamase 课题组系统地研究了 $[W_5O_{18}]^{6-}$ 和稀土离子配位形成的化合物的结构，并讨论了镧系收缩对该系列阴离子结构的影响[27-32]。

1994 年，Burns 等人报道了分子组成为 $K_2Na_6[Pd_2W_{10}O_{36}] \cdot 22H_2O$ 的化合物[33]，在该化合物的阴离子中，$[W_5O_{18}]^{6-}$ 基团仍然作为四齿配体，与稀土配位不同之处在于每个 $[W_5O_{18}]^{6-}$ 基团同时和两个 Pd^{2+} 配位，而每一个 Pd^{2+} 又同时与两个 $[W_5O_{18}]^{6-}$ 基团配位，结构如图 5-15 所示。

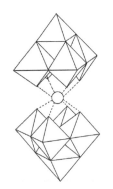

图 5-14 $[W_5O_{18}]^{6-}$ 与稀土离子构成阴离子的多面体

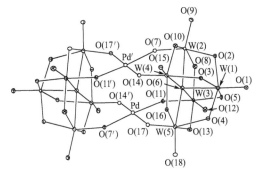

图 5-15 $[Pd_2W_{10}O_{36}]^{8-}$ 原子连接图

1998 年 Naruke 等人报道了分子组成为 $K_{18.5}H_{1.5}[Ce_3(CO_3)(SbW_9O_{33})(W_5O_{18})_3] \cdot 14H_2O$ 的化合物[34]，在该化合物中，3 个 $[W_5O_{18}]^{6-}$ 基团和 1 个 $[SbW_9O_{33}]^{9-}$ 基团相连构成一个笼，$[Ce(CO_3)]^{7-}$ 基团位于笼的中间（图 5-16）。

2007 年，Errington 报道了一系列由锆占据 $[W_5O_{18}]^{6-}$ 中的空位，然后通过有机配体与锆配位构成的新型化合物[35]，在这些化合物中，金属锆原子占据 $[W_5O_{18}]^{6-}$ 中失去 WO 基团的位置，以不同的方式与有机基团通过配位键结合。图 5-17～图 5-23 为一系列化合物的结构。

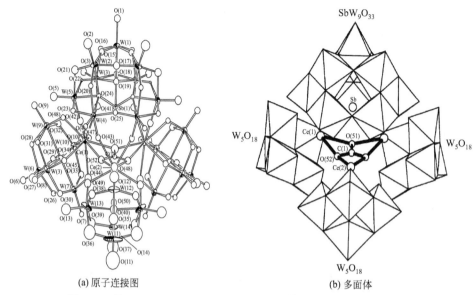

(a) 原子连接图 (b) 多面体

图 5-16 $[Ce_3(CO_3)(SbW_9O_{33})(W_5O_{18})_3]^{20-}$ 原子连接图、多面体

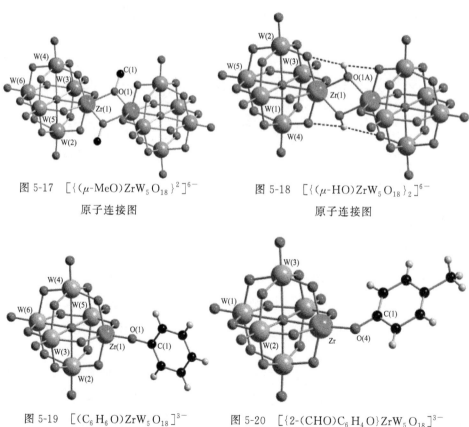

图 5-17 $[\{(\mu\text{-MeO})ZrW_5O_{18}\}_2]^{6-}$ 原子连接图

图 5-18 $[\{(\mu\text{-HO})ZrW_5O_{18}\}_2]^{6-}$ 原子连接图

图 5-19 $[(C_6H_6O)ZrW_5O_{18}]^{3-}$ 原子连接图

图 5-20 $[\{2\text{-}(CHO)C_6H_4O\}ZrW_5O_{18}]^{3-}$ 原子连接图

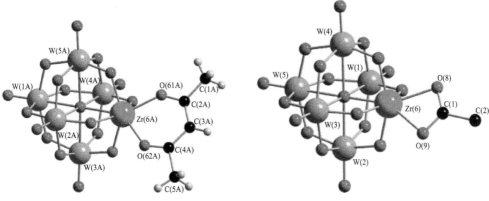

图 5-21　$[(acac)ZrW_5O_{18}]^{3-}$ 原子连接图　　图 5-22　$[(CH_3CO_2)ZrW_5O_{18}]^{3-}$ 原子连接图

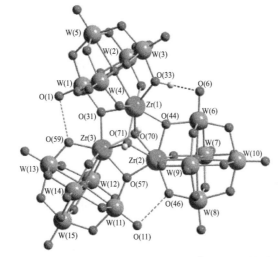

图 5-23　$[(\mu\text{-}OH)_2(ZrW_5O_{18})_3H]^{7-}$ 原子连接图

2009 年，Errington 又报道了两个 $[W_5O_{18}]^{6-}$ 基团和 Co^{2+} 形成的化合物[36]，这两个化合物中，Co^{2+} 占据 WO 基团的位置（图 5-24 和图 5-25）。

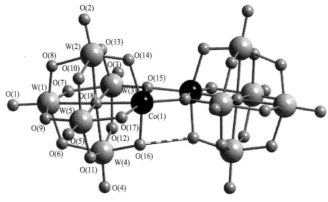

图 5-24　$[\{CoW_5O_{18}H\}_2]^{6-}$ 原子连接图

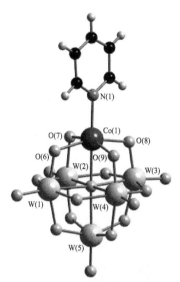

图 5-25 $[(C_5H_5N)CoW_5O_{18}H]^{3-}$ 原子连接图

参考文献

[1] Guillou N, Férey G. Hydrothermal synthesis and structural characterization of two layered diamine pentamolybdates: $(C_2H_{10}N_2)[Mo_5O_{16}]$ and $(C_4H_{12}N_2)[Mo_5O_{16}]$. J Solid State Chem, 1999, 147: 240-246.

[2] Gall P, Gougeon P. Structure of $SrMo_5O_8$ containing chains of bioctahedral Mo_{10} clusters. Acta Cryst, 1994, C50: 7-9.

[3] Red'kin A F, Ponomarev V I, Kostromin N P. Hydrothermal synthesis and properties of sodium pentatungstate monohydrate $Na_2W_5O_{16} \cdot H_2O$. Russian Journal of Inorganic Chemistry, 2000, 45 (8): 1177-1183.

[4] Sronskowski R, Simon A, Mertin W. Synthese and kristallstruktur von $PbMo_5O_8$; ein reduziertes oxomolybdat mit $Mo_{10}O_{28}$-oktaederdoppeln. Z Anorg Allg Chem, 1991, 602: 49-63.

[5] Gougeon P, Gall P, Sergent M. Structure of $GdMo_5O_8$. Acta Cryst, 1991, C47: 421-423.

[6] Lasocha W, Schenk H. Crystal structure of anilinium pentamolybdate from powder diffraction data. The solution of the crystal structure by direct methods package POWSIM. J Appl Cryst, 1997, 30: 909-914.

[7] Gatehouse B M, Miskin B K. The crystal structures of caesium pentamolybdate, $Cs_2Mo_5O_{16}$ and caesium heptamolybdate, $Cs_2Mo_7O_{22}$. Acta Cryst, 1975, B31: 1293-1299.

[8] Wiesmann W, Weitzel H, Svoboda I, Fuess H. The crystal structure of lithiumpentamolybdate $Li_4Mo_5O_{17}$. Zeitschrift für Kristallographie, 1997, 212: 795-800.

[9] Gougeon P, Potel M, Sergent M. Structure of $SnMo_5O_8$ containing bioctahedral Mo_{10}

clusters. Acta Cryst, 1990, C46: 1188-1190.

[10] Zhang Shu-Fang, Sun Yan-Qiong, Yang Guo-Yu. Bis [benzidinium (1-)] pentamolybdate. Acta Cryst, 2004, C60: m299-m301.

[11] Zurkova L, Miklova K. Studium der Reaktionen von $TlNO_3$-V_2O_5 in fester phase. J Thermal Anal, 1975, 8: 577.

[12] Gaplovska K, Zurkova L. Thermal properties of pentavanadates. Journal of Thermal Analyses, 1981, 20: 463-469.

[13] Day V W, Klemperer W G, Yaghi O M. A new structure type in polyoxoanion chemistry: synthesis and structure of the $V_5O_{14}^{3-}$ anion. J Am Chem Soc, 1989, 111: 4518-4519.

[14] Filowitz M, Klemperer W G, Shum W. Synthesis and characterization of the pentamolybdate ion, $Mo_5O_{17}H^{3-}$. Journal of the American Chemical Society, 1978, 100 (8): 2580-2581.

[15] Niu Jingyang, Wang Ziliang, Wang Jingping. Hydrothermal synthesis and crystal structure of a novel 4,4'-bipyridine bridging dumbbell-like dimeric pentamolybdate: [4,4'-H_2bipy]$_3$[4,4'-Hbipy]$_2$[(Mo_5O_{17})$_2$(4,4'-bipy)]. Inorg Chem Commun, 2004, 7: 556-558.

[16] Proust A, Gouzerh P, Robert F. Molybdenum oxo nitrosyl complexes: 1. Defect lindqvist compounds of the type [Mo_5O_{13}(OR)$_4$(NO)]$^{3-}$ (R=CH_3, C_2H_5). Solid-state interactions with alkali-metal cations. Inorg Chem, 1993, 32: 5291-5298.

[17] Villanneau R, Proust A, Robert F, Veillet P, Gouzerh P. Synthesis, structure, and magnetic properties of (n-Bu_4N)$_2$[{Ni(MeOH)$_2$}$_2${Mo(NO)}$_2$(μ_3-OH)$_2$(μ-OMe)$_4${Mo_5O_{13}(OMe)$_4$(NO)}$_2$], a new type of polyoxometalate incorporating a rhomb-like cluster. Inorg Chem, 1999, 38: 4981-4985.

[18] Villanneau R, Proust A, Robert F, Villain F, Verdaguer M, Gouzerh P. Polyoxoanion-supported pentamethylcyclopentadienylrhodium complexes: syntheses and structural characterization by EXAFS. Polyhedron, 2003, 22: 1157-1165.

[19] Villanneau R, Proust A, Robert F, Gouzerh P. Coordination chemistry of the soluble metal oxide analogue [Mo_5O_{13}(OCH_3)$_4$(NO)]$^{3-}$ with manganese carbonyl species. Chem Eur J, 2003, 9: 1982-1990.

[20] Schaefer E. Tungsten componds. Zeitschrift fuer Anorganische Chemie, 1904, 38, 142-183.

[21] Pereira J G. New color reaction of hydrogen gas. Anales de la Real Sociedad Espanola de Fisicay Quimica, 1913, 10: 370-381.

[22] Neu F, Schwing-Weill M J. Piperidine tungtates; photochromism of the anhydrous pentatungstate. Bulletin de la Societe chimique de France, 1968, 12: 4821-4828.

[23] Fuchs J, Palm R, Hartl H. $K_7HW_5O_{19}$ · 10H_2O-a novel isopolyxotungstate (Ⅵ). Angrew chem Int Ed Engl, 1996, 35 (22): 2651-2653.

[24] Peacock R D, Weakley T J R. Heteropolytungstate complexes of the lanthanide ele-

ments. Part Ⅰ. Preparation and reactions. J Chem Soc (A), 1971: 1836-1839.

[25] Peacock R D, Weakley T J R. Heteropolytungstate complexes of the lanthanide elements. Part Ⅱ. Electronic spectra: a metal-ligand charge-transfer transition of cerium (Ⅲ). J Chem Soc (A), 1971: 1937-1940.

[26] Iball J, Low J N, Weakley T J R. Heteropolytungstate complexes of the lanthanid elements. Part Ⅲ. Crystal structure of sodium decatungstocerate (Ⅳ) -Water (1/30). J C S Dalton, 1974, 18 (18): 2021-2024.

[27] Ozeki T, Takahashi M, Yamase T. Structure of $K_3Na_4H_2[TbW_{10}O_{36}] \cdot 20H_2O$. Acta Cryst, 1992, C48: 1370.

[28] Yamase T, Ozeki T. Structure of $K_3Na_4H_2[GdW_{10}O_{36}] \cdot 21H_2O$. Acta Cryst, 1993, C49: 1577.

[29] Ozeki T, Yamase T. Structure of $K_3Na_4H_2[SmW_{10}O_{36}] \cdot 22H_2O$. Acta Cryst, 1993, C49: 1574.

[30] Yamase T, Ozeki T, Ueda K. Structure of $NaSr_4[EuW_{10}O_{36}] \cdot 34.5H_2O$. Acta Cryst, 1993, C49: 1572.

[31] Yamase T, Ozeki T, Tosaka M. Octasodium hydrogen decatungstogadolinate triacontahydrate. Acta Cryst, 1994, C50: 1849.

[32] Ozeki T, Yamase T. Hexasodium trihydrogen decatungstosamarate octacosahydrate. Acta Cryst, 1994, C50: 327.

[33] Angus-Dunne S J, Burns R C, Craig D C, Lawrance G A. A novel heteropolymetalate containing palladium (Ⅱ): synthesis and crystal structure of $K_2Na_6[Pd_2W_{10}O_{36}] \cdot 22H_2O$. J Chem Soc Chem Commun, 1994, 523-524.

[34] Naruke H, Yamase T. Crystal structure of $K_{18.5}H_{1.5}[Ce_3(CO_3)(SbW_9O_{33})(W_5O_{18})_3] \cdot 14H_2O$. J Alloys and Compounds, 1998, 268 (1-2): 100-106.

[35] Errington R J, Petkar S S, Middleton P S, McFarlane W, Clegg W, Coxall R A, Harrington R W. Synthesis and reactivity of the methoxozirconium pentatungstate $(^nBu_4N)_6[\{(\mu\text{-MeO})ZrW_5O_{18}\}_2]$: insights into proton-transfer reactions, solution dynamics, and assembly of $\{ZrW_5O_{18}\}^{2-}$ building blocks. J Am Chem Soc, 2007, 129: 12181-12196.

[36] Errington R J, Harle G, Clegg W, Harrington R W. Extending the lindqvist family to late 3d transition metals: a rational entry to CoW_5 hexametalate Chemistry. Eur J Inorg Chem, 2009, 34: 5240-5246.

第 6 章

六金属氧簇

六金属氧簇是一类稳定性较高的金属氧簇,能形成六金属氧簇的元素相对较多,已发现 V、Nb、Ta、Mo、W 等均可形成六金属氧簇,是目前金属种类最多的一种聚金属氧簇阴离子。1952 年,Lindqvist 在研究铌化合物时,首次阐明了六铌氧簇 $[Nb_6O_{19}]^{8-}$ 的结构[1],因此,现在这类结构被称为 Lindqvist 结构。

6.1 六钒氧簇

有关水溶液中 VO_4^{3-} 的存在状态的研究已有相当长的历史和相当多的报道[2-8]。早在 1903 年,Dullberg 等人就提出在溶液中存在 $HV_6O_{17}^{3-}$ 六钒氧簇离子[9],而且该观点得到了很多人的支持,但直到现在,还没有得到具有独立结构的六钒氧簇阴离子形成的化合物。尽管具有独立结构的六钒氧簇阴离子形成的化合物至今没有分离出来,但含有六钒氧簇 $[V_6O_{19}]$ 基团的化合物已有多例报道。

6.1.1 合成

1989 年 Klemperer 等人以 $[(C_5Me_5)Rh(OH)_2]$ 和 V_2O_5 为原料,在水溶

液中反应后，得到一种棕黑色的沉淀，将该沉淀在乙醇和乙腈的混合溶剂中重结晶后，得到了组成为 $[(C_5Me_5)Rh]_4(V_6O_{19}) \cdot 3CH_3CN \cdot CH_3OH \cdot 3H_2O$ 的化合物[10]。该化合物 Isobe 等人在 1988 年的会议论文摘要中曾经报道过，后来又作了详细报道[11]。

6.1.2 结构

$[(C_5Me_5)Rh]_4(V_6O_{19})$ 结构如图 6-1 所示，该化合物为电中性，具有独立结构。化合物由 1 个六钒氧簇 $[V_6O_{19}]^{8-}$ 基团和 4 个与其相连的 $(\eta^5\text{-}C_5Me_5)Rh^{2+}$ 构成。$[V_6O_{19}]^{8-}$ 核具有 Lindqvist 结构，其中的六个金属原子所处环境相同，氧原子可分为三类（图 6-2）：①只与一个金属原子相连的端氧，称为 O_t；②与两个金属原子相连的桥氧，称为 O_b；③位于阴离子中心与 6 个金属原子相连接的桥氧，称为 O_c。整个阴离子由六配位的金属氧（VO_6）八面体通过共棱相连。六钒氧簇核 $[V_6O_{19}]^{8-}$ 在不同化合物中的键长数据列于表 6-1。

图 6-1 $[(C_5Me_5)Rh]_4(V_6O_{19})$ 结构

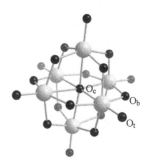

图 6-2 Lindqvist 结构

表 6-1 $[V_6O_{19}]^{8-}$ 在不同化合物中的 V—O 键长　　单位：nm

化合物①	V—O_t	V—O_b	V—O_c	文献
$[(C_5Me_5)Rh]_4(V_6O_{19})$	0.1604	0.1916	0.2246	[10]
$(n\text{-}But)_4N]_2[V_6O_{13}\{O_2NC(CH_2O)_3\}_2]$	0.1602	0.1819 0.2028(烷氧基)	0.2244	[12]
$[(RhCp^*)_4V_6O_{19}]$	0.161	0.193	0.225	[13]
$[(RhCp^*)_3Ir(Cp^*)V_6O_{19}]$	0.161	0.190	0.247	[13]
$[(n\text{-}C_4H_9)_4N]_2[V_6O_{16}][(OCH_2)_3CNHC(O)CHCH_2]_3$	0.160	0.1824 0.2016(烷氧基)	0.2243	[14]
$[C_5H_5NH]_2[V_6O_{13}\{(OCH_2)_3CCH_3\}_2]$	0.1605	0.1886 0.2503(烷氧基)	0.2243	[14]
$[(n\text{-}C_4H_9)_4N]_2[V_6O_{10}(OH)_3\{(OCH_2)_3CNO\}_2]$	0.1593	0.186 0.194(羟基) 0.2017(烷氧基)	0.228	[14]

① 略去溶剂分子。

1991 年，Isobe 报道了分子组成为 $[(Cp^*Ir)_4V_6O_{19}]$ 和 $[(Cp^*Rh)_{4-n}(Cp^*Ir)_nV_6O_{19}]$ 的化合物[12]，同时还再次研究了化合物 $[(Cp^*Rh)_4V_6O_{19}]$，这些化合物均和 Kelemperer 报道的 $[(C_5Me_5)Rh]_4[V_6O_{19}]\cdot 3CH_3CN\cdot CH_3OH\cdot 3H_2O$ 同构。

1990 年，Zubieta 报道了以三羟甲基硝基甲烷和 $[(n-C_4H_9)_4NH_3V_{10}O_{28}]$ 为原料，在氮气气氛下操作，在乙腈溶液中回流，得到了分子组成为 $[(n-C_4H_9)_4N]_2[V_6O_{13}\{O_2NC(CH_2O)_3\}_2]$ 的化合物[13]，在该化合物中，三羟甲基硝基甲烷中的三个羟基氧原子作为桥氧原子参与了阴离子的构成，其阴离子结构如图 6-3 所示。1992 年，Zubieta 以 $[(n-C_4H_9)_4NH_3V_{10}O_{28}]$ 和三羟甲基甲烷衍生物为原料，合成了一系列多金属氧簇核为 $[V_6O_{19}]$ 的化合物[14]，这些化合物阴离子均与 $[V_6O_{13}\{O_2NC(CH_2O)_3\}_2]^{2-}$ 同构。

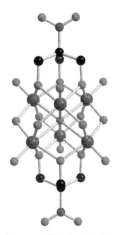

图 6-3 $[V_6O_{13}\{O_2NC(CH_2O)_3\}_2]^{2-}$ 结构

1992 年，Zubieta 以 $[(C_4H_9)_4N]_3[H_3V_{10}O_{28}]$ 和 $(HOCH_2)_3CR$（R = NO_2，CH_2OH 和 CH_3）为原料，在乙腈溶液中反应，得到了分子组成为 $[(C_4H_9)_4N]_2[V_6O_{13}\{(OCH_2)_3CR\}_2]$（R= NO_2，CH_2OH，CH_3）的化合物，用有机肼还原 $[(C_4H_9)_4N]_2[V_6O_{13}\{(OCH_2)_3CCH_3\}_2]$，可得到混合价化合物 $[(C_4H_9)_4N]_2[V_4^{IV}V_2^VO_9(OH)_4\{(OCH_2)_3CCH_3\}_2]$ 和 $[(C_4H_9)_4N]_2[V_6^{IV}O_7(OH)_6\{(OCH_2)_3CCH_3\}_2]\cdot 2CH_2Cl_2\cdot 0.5C_6H_5NNC_6H_5$，其中质子的位置由单晶 X 射线确定，质子化和还原没有关联，化合物 $[(C_4H_9)_4N]_2[V_6O_{13}\{(OCH_2)_3CCH_3\}_2]$ 与 $HBF_4\cdot O(C_2H_5)$ 反应，可得到化合物 $[V_6O_{11}(OH)_2\{CH_3C(CH_2O)_3\}]$，其中的两个氧发生了质子化，质子化的位置由单晶 X 射线确定，但分子内的钒并未被还原，仍为六价[15]。次年，Zubieta 又以 $[Bu_4N]_3V_5O_{14}$ 和 $RC(CH_2OH)_3$ 为原料，在甲醇溶液中反应，在得到分子组成为 $[Bu_4N]_2[V_6O_{13}\{(OMe)_3(OCH_2)_3CR\}]$（R= Et，—$CH_2OH$）的化合物的同时，还得到了组成为 $[Bu_4N]_2[V_6O_8\{(OCH_2)_3CR\}_2\{(OCH_2)_2C(CH_2OH)R\}_4]$ 的化合物[16]，在化合物 $[Bu_4N]_2[V_6O_{13}\{(OMe)_3(OCH_2)_3CR\}]$ 中，其核心结构与前述 $[V_6O_{19}]$ 同构，而在化合物 $[Bu_4N]_2[V_6O_8\{(OCH_2)_3CR\}_2\{(OCH_2)_2C(CH_2OH)R\}_4]$ 中，其核心结构则是一个 $[V_4O_{16}]$ 单元与两个钒氧四方锥共棱相连

图 6-4 $[V_6O_8\{(OCH_2)_3CR\}_2\{(OCH_2)_2C(CH_2OH)R\}_4]^{2-}$ 中钒氧多面体连接图

（图 6-4）。

Muller 报道，以偏钒酸钠或偏钒酸铵在一定的条件下与季戊四醇反应，根据条件的不同，可分别得到季戊四醇位于顺式和反式的异构体：$cis\text{-}Na_2[V_6^{IV}O_7(OH)_6\{(OCH_2)_3CCH_2OH\}_2] \cdot 8H_2O$，$cis\text{-}[CN_3H_6]_3[V^{IV}V_5^VO_{13}\{(OCH_2)_3CCH_2OH\}] \cdot 4.5H_2O$ 和 $trans\text{-}(CN_3H_6)_2[V_6^VO_{13}\{(OCH_2)_3CCH_2OH\}_2] \cdot H_2O^{[17]}$，结构如图 6-5 所示。

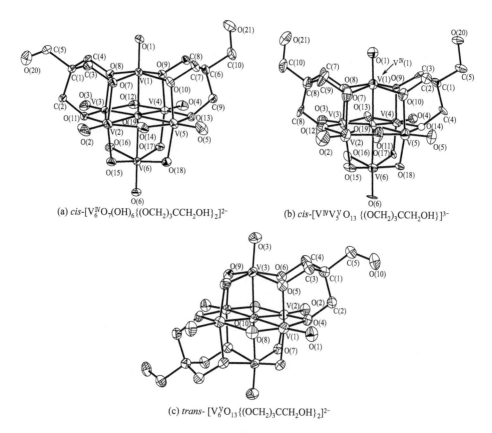

(a) $cis\text{-}[V_6^{IV}O_7(OH)_6\{(OCH_2)_3CCH_2OH\}_2]^{2-}$

(b) $cis\text{-}[V^{IV}V_5^VO_{13}\{(OCH_2)_3CCH_2OH\}]^{3-}$

(c) $trans\text{-}[V_6^VO_{13}\{(OCH_2)_3CCH_2OH\}_2]^{2-}$

图 6-5 化合物中阴离子结构

Kessler 发现，$VO(OEt)_3$ 在放置的时候，可以发生微水解，生成黄绿色晶体，结构分析表明，该晶体分子组成为 $V_6O_7(OEt)_{12}^{[18]}$，其核心为 $[V_6O_{19}]$，12 个乙氧基中的氧原子作为桥氧原子分别连接两个钒原子，结构如图 6-6 所示。

由 $NaVO_3 \cdot 2H_2O$ 和邻菲啰啉在 pH=8.5、150℃水热条件下反应，冷却后用甲醇覆膜生长晶体，得到了分子组成为 $[(phen)_4V_6O_{12}(CH_3OH)_4] \cdot 2CH_3OH \cdot 4H_2O$ 的化合物[19]，在该化合物中 6 个钒原子几乎位于同一个平面内，4 个邻菲啰啉分子和 4 个甲醇分子参与了钒的配位，结构如图 6-7 所示。

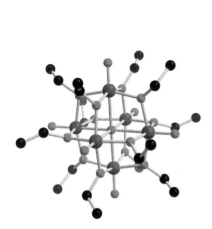

图 6-6　$V_6O_7(OEt)_{12}$ 原子连接图

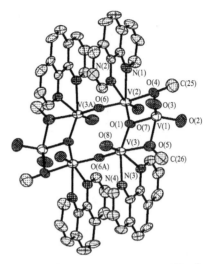

图 6-7　$[(phen)_4V_6O_{12}(CH_3OH)_4]$
分子原子连接图

Krebs 用二 [2-(1-甲基咪唑基)]-4-甲氧基苯基甲醇 (bmimpm)、三乙胺和 [VO(acac)] (acac= 乙酰丙酮) 在甲醇溶液中反应, 得到了分子组成为 [{VO(bmimpm)(acac)}$_2${V$_6$O$_{13}$(OCH$_3$)$_6$}] 的化合物[20], 阴离子中的 6 个甲氧基均匀地分布在阴离子上。Hayashi 研究组用 [(C$_2$H$_5$)$_4$N] VO$_3$ 和 [Pd(C$_6$H$_5$CN)$_2$Cl$_2$] 在乙腈中反应, 得到了分子组成为 [(C$_2$H$_5$)$_4$N]$_4$[PdV$_6$O$_{18}$] 的化合物[21], 在该化合物阴离子中, 6 个四配位的钒氧四面体构成一个六元环, 钯原子位于环的中央, 通过和 4 个氧原子配位与环相连, 结构如图 6-8 所示。

Hill 以 (HOCH$_2$)$_3$CNHCO(4-C$_5$H$_4$N) 和 [(n-C$_4$H$_9$)$_4$N]$_2$[H$_3$V$_{10}$O$_{28}$] 为原料, 制得了分子组成为 [(n-C$_4$H$_9$)$_4$N]$_2$ [V$_6$O$_{13}${(CH$_2$O)$_3$CNHCO(4-C$_5$H$_4$N)}$_2$] · 1.8DMF 的化合物, 并以此为原料和相应的过渡金属硝酸盐反应制得了一系列配位聚合物 [M(H$_2$O)$_2$(DMF)$_2$(V$_6$O$_{13}${(CH$_2$O)$_3$CNHCO(4-C$_5$H$_4$N)}$_2$)]$_n$ (M=Co, Mn, Ni, Zn)[22]。环丙沙星与偏钒酸铵在水热条件下反应, 得到了分子组成为 V$_4$O$_{10}$(μ-O)$_2$[VO(H-Ciprof)$_2$]$_2$·13H$_2$O 的化合物[23], 在该化合物中 4 个四配位钒氧四面体通过共顶点相连形成一个环, 两个六配位的钒氧八面体通过共顶点和四元环相连, 结构如图 6-9 所示。

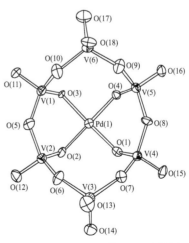

图 6-8　$[PdV_6O_{18}]^{2-}$ 原子连接图

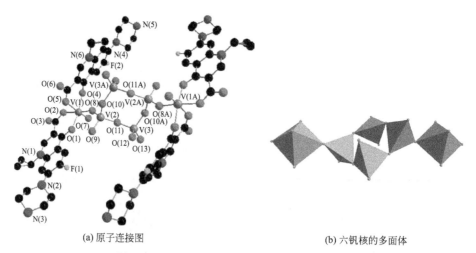

(a) 原子连接图　　　　　　　　(b) 六钒核的多面体

图 6-9　$V_4O_{10}(\mu\text{-}O)_2[VO(H\text{-}Ciprof)_2]_2$ 结构

$[(n\text{-}C_4H_9)_4N]_2[H_3V_{10}O_{28}]$ 和 4-[(三羟甲基甲氨基)甲基]苯甲酸(4-$HOOCC_6H_4)CH_2NHC(CH_2OH)_3$ 在二甲基乙酰胺中反应,可得到分子组成为 $[(n\text{-}C_4H_9)_4N]_2[V_6O_{13}\{(CH_2O)_3CNHCH_2C_6H_4\text{-}4\text{-}COOH\}_2]$ 的化合物,并以该化合物和 $Tb(NO_3)_3$ 为原料,利用扩散的方法,得到了化合物 $[Tb(bpdo)_2\{[V_6O_{13}(CH_2O)_3C(NHCH_2C_6H_4\text{-}4\text{-}COOH)_2]\}] \cdot 1.5DMF \cdot 3.0EG$ (bpdo = N,N'-二氧化-4,4'-联吡啶;EG=乙二醇)[24],在该化合物中,三羟基中的氧作为 $[V_6]$ 中的桥氧参与阴离子的构成,位于苯环对位的羧基与稀土离子配位,化合物具有二维层状结构。

4,4'-二甲基-2,2'-联吡啶和硫酸氧钒 $VOSO_4 \cdot 3H_2O$ 在水和甲醇的混合溶剂中反应,可生成化合物 $[V_6O_{12}(OCH_3)_4(dmbpy)_4] \cdot 2H_2O$ (dmbpy = 4,4'-二甲基-2,2'-联吡啶)[25],该化合物和前述文献[19]报道的化合物 $[(phen)_4V_6O_{12}(CH_3OH)_4] \cdot 2CH_3OH \cdot 4H_2O$ 同构。

硫酸氧钒与对-叔丁基杯(4)芳烃(calyx)在厌氧条件下于甲醇溶液中进行溶剂热反应,通过改变有机阳离子,得到了一系列分子组成为(cat)$[V_6O_6(OCH_3)_8(calix)(CH_3OH)]$ (cat = Et_4N^+, NH_4^+, PyH^+, Et_3NH^+)的化合物[26],化合物阴离子结构如图 6-10

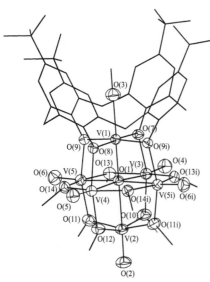

图 6-10　$[V_6O_6(OCH_3)_8(calix)(CH_3OH)]^-$ 结构

所示，对-叔丁基杯（4）芳烃中的 4 个羟基氧作为 [V_6] 阴离子中的桥氧参与阴离子的形成。化合物在生产的过程中发生了钒的还原，分子内存在 V(Ⅲ)/V(Ⅳ) 作用，表现出铁磁性。

(Bu_4N)$_3$[$H_3V_{10}O_{28}$] 和 N-三羟甲基甲基氨基甲酰基二茂铁反应可得到分子组成为 (Bu_4N)$_2$[{FeC(O)NHC(CH$_2$O)$_3$}$_2$V$_6$O$_{13}$] 的化合物[27]，在该化合物中 N-三羟甲基甲基氨基甲酰基二茂铁中的三个羟基作为 [V_6] 阴离子中共面的三个桥氧参与阴离子的构成，结构如图 6-11 所示。

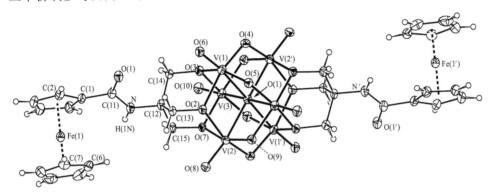

图 6-11　[{FeC(O)NHC(CH$_2$O)$_3$}$_2$V$_6$O$_{13}$] 原子连接图

室温下用 H 形管缓慢将 [VO(acac)$_2$](acac＝乙酰丙酮) 和 1,10-邻菲啰啉（或 2,2'-联吡啶）的甲醇溶液扩散到溶有焦磷酸钠 Na$_4$P$_2$O$_7$ 的水溶液中，分别得到了组成为 [V$_6$O$_{12}$(OCH$_3$)$_4$(L)$_4$]·2CH$_3$OH·4H$_2$O [L＝1,10-邻菲啰啉（phen）] 和 [V$_6$O$_{12}$(OCH$_3$)$_4$(L)$_4$]·4H$_2$O [L＝2,2'-bipyridine(bpy)] 的化合物[28]，在这两个化合物中阴离子 [V_6] 核心不是 Lindqvist 结构，而是与文献报道的 [(phen)$_4$V$_6$O$_{12}$(CH$_3$OH)$_4$]·2CH$_3$OH·4H$_2$O 及 [V$_6$O$_{12}$(OCH$_3$)$_4$(dmbpy)$_4$·2H$_2$O] 同构[19,25]。

由季戊四醇参与形成的化合物 trans-(CN$_3$H$_6$)$_2$[V$_6^V$O$_{13}${(OCH$_2$)$_3$CCH$_2$OH}$_2$]·H$_2$O[17]，由于分子内含有两个羟基，在 4-(N,N-二甲氨基) 吡啶催化下可与酸酐发生酯化反应形成新的衍生物[29]。在催化剂的存在下，也可与硬脂酸发生酯化反应，生成组成为 (Bu$_4$N)$_2$[V$_6$O$_{13}${(OCH$_2$)$_3$CCH$_2$OOC(CH$_2$)$_{16}$CH$_3$}$_2$] 的化合物[30]，其阴离子结构如图 6-12 所示。

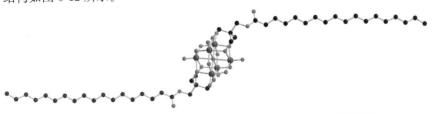

图 6-12　[V$_6$O$_{13}${(OCH$_2$)$_3$CCH$_2$OOC(CH$_2$)$_{16}$CH$_3$}$_2$]$^{2-}$ 结构

TBA$_3$[H$_3$V$_{10}$O$_{28}$] 和烟酰基 N-2{(2-羟甲基)-1,3-丙二醇} 反应可以得到分子式为 TBA$_2$[V$_6$O$_{13}${(OCH$_2$)$_3$CNHC(O)-3-C$_5$H$_4$N}$_2$] 的化合物[31]，在化合物中，两个有机配体位于反式，分别用三个羟基作为桥氧参与阴离子的构成，结构如图 6-13 所示。

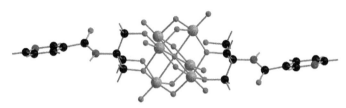

图 6-13 [V$_6$O$_{13}${(OCH$_2$)$_3$CNHC(O)-3-C$_5$H$_4$N}$_2$]$^{2-}$ 结构

化合物阴离子中吡啶基上的氮原子具有较强的配位能力，可与过渡金属配位形成新颖结构的化合物，与 PdCl$_2$(CH$_3$CN)$_2$ 反应可以形成结构如图 6-14 所示的化合物。

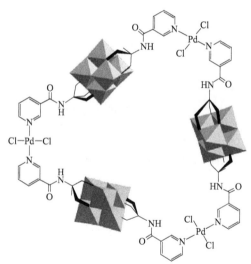

图 6-14 [V$_6$O$_{13}${(OCH$_2$)$_3$CNHC(O)-3-C$_5$H$_4$N}$_2$]$^{2-}$ 通过与 Pd^{2+} 配位构成的三聚结构

6.2 六铌氧簇

自 1952 年 Lindqvist 报道六铌氧簇阴离子 [Nb$_6$O$_{19}$]$^{8-}$ 的结构以后[1]，又有很多 [Nb$_6$O$_{19}$]$^{8-}$ 和不同抗衡离子形成化合物的报道，特别是法国化学家，在该领域做了大量的研究工作[32-39]，但由于 Nb 本身的惰性，难以形成新型化合物，对该类化合物的研究主要集中于由 [Nb$_6$O$_{19}$]$^{8-}$ 形成的盐及其在水溶液中的性质，而由 [Nb$_6$O$_{19}$]$^{8-}$ 形成的相关化合物，则报道较少。

6.2.1 合成

六铌氧簇钾盐（通称铌酸钾）$K_7HNb_6O_{19} \cdot nH_2O$ 可由五氧化二铌 Nb_2O_5 与氢氧化钾 KOH 固体在镍坩埚中熔融，然后用少量水萃取除去过量的 KOH。粗铌酸钾用乙醇-水混合溶剂洗涤，然后溶于水中，过滤，向滤液中加入乙醇，使铌酸钾沉淀下来，过滤，分别用体积比为 1:1 的乙醇-水混合溶剂洗涤、95% 乙醇和乙醚洗涤，然后空气干燥，即可得到产品[40,41]。六铌氧簇钠盐（铌酸钠）$Na_7HNb_6O_{19} \cdot nH_2O$ 可用类似的方法制得。

6.2.2 谱学表征

(1) 红外光谱

六铌氧簇阴离子 $[Nb_6O_{19}]^{8-}$ 的红外光谱在 $400 \sim 900 cm^{-1}$ 范围内表现出 $Nb=O$ 和 $Nb-O-Nb$ 的伸缩振动吸收峰[42-44]。由于阴离子质子化程度的不同，红外光谱中表现出一定的差别，表 6-2 列出了不同质子化的六铌氧簇钾盐的红外光谱数据。

表 6-2 $K_{8-x}[H_xNb_6O_{19}] \cdot nH_2O$ 的红外光谱数据　　单位：cm^{-1}

化合物	$K_7[HNb_6O_{19}] \cdot 10H_2O$	$K_6[H_2Nb_6O_{19}] \cdot 13H_2O$	$K_5[H_3Nb_6O_{19}] \cdot 10H_2O$
$\nu(Nb-O_t)$	858vs	860vs 842s	860vs 840sh
$\nu(Nb-O_b)_{asym}$	770sh 670vs	770sh 670vs	780m 663vs
$\nu(Nb-O_b)_{sym}$	526vs	520vs	525vs
$\nu(Nb-O_c)$	414s	406s	414s

注：O_t—端氧；O_b—桥氧；O_c—中心氧。

(2) 核磁

Alam 等人研究了 $Na_7[HNb_6O_{19}] \cdot 15H_2O$ 在溶液中和固体的核磁，如表 6-3 所示，确定了六铌氧簇内氧及羟基的环境[45]。

表 6-3 $Na_7[HNb_6O_{19}] \cdot 15H_2O$ 的溶液和固体 ^{17}O 核磁[45]

样品	化学位移(半峰宽/Hz)				
	O_t	O_b	NbOH	O_c	H_2O
溶液(℃)					
20①	631(260)	376(1720)	186(250)	34(100)	0(170)
30①	633(230)	375(1330)	185(230)	35(100)	0(170)
50①	633(225)	375(620)	186(180)	36(120)	−1(150)
20②	633(225)	372(1350)	182(225)	34(100)	0(170)
20③	630(225)	376(1000)	186(250)	33(120)	0(180)
30③	633(250)	375(800)	185(250)	35(100)	0(150)
50③	632(300)	375(650)	187(215)	36(150)	−1(180)

续表

样品	化学位移(半峰宽/Hz)				
	O_t	O_b	NbOH	O_c	H_2O
固体(℃)					
25 (54.28 MHz)	631(1100) 575(1175)	386(1450) (426)	226(700)	31(330)	18(275) [−90(9000)]
25 (81.40 MHz)	636(1600) 576(1700)	386(2100) (423)	227(1500)	31(500)	19(800)
25 (81.40 MHz)	636(1600) 576(1700)	386(2100) (423)	227(1500)	31(500)	19(800)

① 样品溶于磷酸缓冲溶液，pH=7.4。
② 样品溶于碳酸钠缓冲溶液，pH=10.0。
③ 样品溶于水，最终 pH=11.0。

结果显示，无论是在溶液中，还是在固态，六铌氧簇 Lindqvist 阴离子上连接的质子都位于桥氧上，表明桥氧是阴离子中碱性最强的氧位置。溶液中，阴离子中的氧同溶剂中的氧存在快速交换，阴离子中不同位置的氧交换的速度顺序是 $O_b > O_t >$ NbOH $\gg O_c$，O_t 和 NbOH 的交换速度随 pH 降低而加快，而 O_b 的交换速度受 pH 影响很小，表明 pH 控制的 $[Nb_6O_{19}]^{8-}$ 的分解涉及 O_t 和 NbOH 的质子化。研究还表明，NbOH 基团在溶液中可以在从中性到碱性的宽范围（pH=7~11.0）内存在，比较 ^{17}O 的溶液和固态 NMR 数据，可以发现在溶液中质子与桥氧的结合更强，而正是由于质子的结合，降低了阴离子簇的总电荷，从而使阴离子簇稳定。

6.2.3 结构

1952 年，Lindqvist 通过对 $Na_7HNb_6O_{19} \cdot 16H_2O$ 的研究，首次描述了六铌氧簇阴离子的结构[1]，但在他给出的结构分析中并没有确定氧原子的位置。1980 年，Goiffon 首次报道了该阴离子的精确结构[35]。如图 6-15 所示，阴离子由 6 个 NbO_6 八面体通过共边紧密堆积而成，阴离子中铌原子所处环境相同，氧原子可分为 3 类（图 6-2）：O_t 只与一个铌原子相连的端氧，共 6 个；O_b 与两个铌原子相连的桥氧，共 12 个；O_c 与六个铌原子相连，位于阴离子中心的中心氧，只有 1 个。晶体结构数据测定表明，Nb—O_c 键长为 2.3~2.5Å，Nb—O_t 键长为 1.75~1.80Å，桥氧键 Nb—O_b 为 2.0Å 左右。由于六铌氧簇的桥氧易于质子化，对每一个 NbO_6 八面体，质子化的桥氧形成的 Nb—O_b 键长由 2.0Å 增加到 2.1Å 左

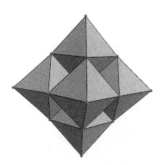

图 6-15 Lindqvist 结构多面体

右,相应的,位于 $HNbO_6$ 八面体中的其他氧原子形成的 $Nb—O_b$ 键将缩短到 1.90~1.95Å[46]。

6.2.4 相关化合物

由于六铌氧簇阴离子只有在碱性条件下才能稳定存在,因此由六铌氧簇阴离子参与形成的化合物报道不多,除了与碱金属形成的盐外,只有少量的由六铌氧簇参与形成的化合物的报道。

1967 年,Pope 报道由铌酸钾与硫酸锰在过硫酸铵存在下反应制得了分子组成可能是 $K_{10}H_{2n}MnNb_{12}O_{37+n}·(22-n)H_2O$ 的化合物[47],但未能得到化合物的晶体结构。1969 年,Flynn 报道铌酸钠与反式 $[Co(en)_2Cl_2]Cl$ 和 $Cr(en)_3I_3·H_2O$ 反应可分别制得分子组成为 $Na_5Co(C_2H_8N_2)Nb_6O_{19}·18H_2O$ 和 $Na_5Cr(C_2H_8N_2)Nb_6O_{19}·18H_2O$ 的化合物[48],随后又报道了铌酸钠分别与 $Mn(OOCCH_3)_2·4H_2O$ 和 $Ni_2SO_4·7H_2O$ 反应制得分子组成为 $Na_{12}MnNb_{12}O_{38}·48H_2O$ 和 $Na_{12}NiNb_{12}O_{38}·50H_2O$ 的化合物[49],同样这些报道没有确定化合物的结构。稍后,Flynn 用 X 射线单晶衍射的方法确定了化合物 $Na_{12}MnNb_{12}O_{38}·50H_2O$ 的晶体结构[50],化合物阴离子结构如图 6-16 所示,每个六铌氧簇阴离子通过三个桥氧原子与锰离子配位,形成夹心型结构阴离子。

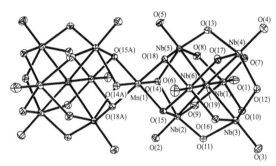

图 6-16 $[MnNb_{12}O_{38}]^{12-}$ 结构

1994 年,Yamase 等人报道了分子组成为 $Na_7H_{19}[\{Eu_3O(OH)_3(OH_2)_3\}_2Al_2(Nb_6O_{19})_5]·47H_2O$ 的含六铌氧簇的化合物[51],1996 年,该研究组又报道了同构的 $Na_8H_{18}[\{Er_3O(OH)_3(H_2O)_3\}_2Al_2(Nb_6O_{19})_5]·40.5H_2O$[52],该化合物阴离子结构如图 6-17 所示,阴离子由 2 个 $[Ln_3O(OH)_3(H_2O)_3]^{4+}$ 簇、2 个 Al^{3+} 和 5 个六铌氧簇阴离子组成。1999 年他们又报道了一个混合稀土的化合物 $Na_7H_{19}[Tb_{4.3}Eu_{1.7}O_2(OH)_6(H_2O)_6Al_2(Nb_6O_{19})_5]·47H_2O$[53]。

2000 年,Cavaleiro 报道了分子组成为 $[Cr_2(\mu-OH)(H_3Nb_6O_{19})(en)_3$

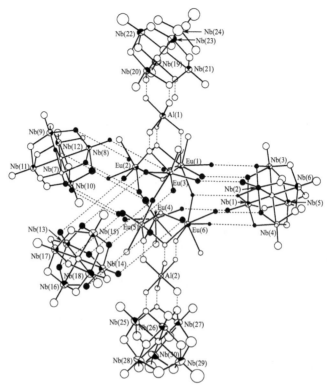

图 6-17 $[\{Eu_3O(OH)_3(H_2O)_3\}_2Al_2(Nb_6O_{19})_5]^{26-}$ 结构

(H_2O)] 的化合物[54],但由于该报道为非晶态化合物,没有准确的结构数据。

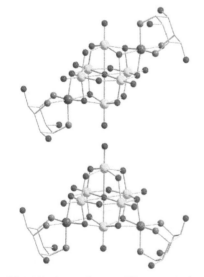

图 6-18 $\{trans\text{-}[Nb_6O_{19}][Ni(taci)]_2\}^{4-}$ 和$\{cis\text{-}[H_2Nb_6O_{19}][Ni(taci)]_2\}^{2-}$ 结构

Hegetschweiler 等人在 2002 年报道了由溶剂热法制得的分子组成为 $[Ni(taci)_2]_2\{trans\text{-}[Nb_6O_{19}][Ni(taci)]_2\}$ 和 $[Na(H_2O)_6]_2\{cis\text{-}[H_2Nb_6O_{19}][Ni(taci)]_2\}\cdot 18H_2O$ 的化合物[55],结构如图 6-18 所示。

由图中可知,六铌氧簇阴离子通过 3 个相邻的桥氧与 2 个 $[Ni(taci)]^{2+}$ 配离子连接[29]。最近,Nyman 研究组报道了多个由六铌氧簇阴离子与 $[CuL_x]$(L= en,NH_3,H_2O)形成的化合物[56],这些化合物表现出六铌氧簇阴离子与 $[CuL_x]$ 配离子配位时的多样性。图 6-19~图 6-23 描述了这些化合物中阴离子的结构。

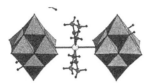

图 6-19　$[(Nb_6O_{19}H_2)-Cu(en)_2-(Nb_6O_{19}H_2)]^{10-}$ 链状结构

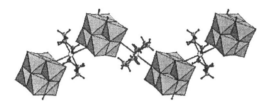

图 6-20　$[-(Nb_6O_{19}H_2)-Cu(en)_2-]^{4-}$ 链状结构

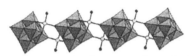

图 6-21　$\{-[Nb_6O_{19}]-[Cu(NH_3)_2]-\}^{4-}$ 链状结构

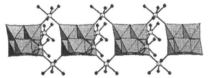

图 6-22　$\{-[Nb_6O_{19}][Cu(NH_3)_2(H_2O)]_2-[Cu(H_2O)_4]_2-\}$ 链状结构

2008 年，我们实验室报道了一个分子组成为 $\{Nb_6O_{19}[Cu(2,2'\text{-bpy})]_2[Cu(2,2'\text{-bpy})_2]_2\} \cdot 19H_2O$ 的化合物，在该化合物中，$[Nb_6O_{19}]^{8-}$ 以位于同一平面的三个桥氧和端氧两种配位方式与铜配位，$4,4'$-联吡啶与铜剩余的配位位置配位[57]，结构如图 6-24 所示。

铌酸钾 $K_7HNb_6O_{19} \cdot 13H_2O$ 在水热条件下与 $Re(CO)_5Br$ 反应，可以得到三羰基铼阳离子 $[Re(CO)_3]^+$ 通过 $[Nb_6O_{19}]^{8-}$ 中的三个共面的桥氧原子与阴离子结合的化合物 $K_7[Re(CO)_3Nb_6O_{19}]$。$K_7HNb_6O_{19} \cdot 13H_2O$ 与 $[Re(CO)_3(CH_3CN)_3]ClO_4$ 或 $[Mn(CO)_3(CH_3CN)_3]ClO_4$ 反应，$[Nb_6O_{19}]^{8-}$ 可以通过两组共面的桥氧原子与两个 $[Re(CO)_3]^+$ 结合，形成顺式和反式结构的阴离子，阴离子结构如图 6-25 所示，产物中每种异构体的含量与反应的温度相关[58]。

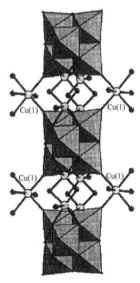

图 6-23　$\{-[Nb_6O_{19}][Cu(NH_3)_2(H_2O)]_2[Cu(H_2O)_4]_2-\}$ 链状结构

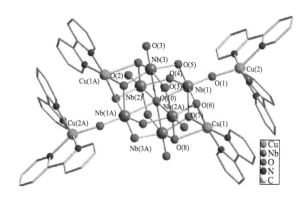

图 6-24　{Nb_6O_{19}[Cu(2,2′-bpy)]$_2$[Cu(2,2′-bpy)$_2$]$_2$}·19H_2O 结构

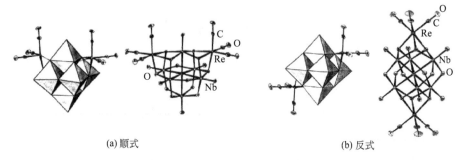

(a) 顺式　　　　　　　　　(b) 反式

图 6-25　顺式和反式 [Nb_6O_{19}{Re(CO)$_3$}$_2$]$^{6-}$ 结构

6.3　六钽氧簇

1954 年 Lindqvist 和 Aronsson 首次报道了化合物 $K_8Ta_6O_{19}$·16H_2O 中六钽氧簇阴离子的结构表征[59]。钽酸盐的溶液化学和铌酸盐的溶液化学几乎同时开展，但由于钽的惰性更强，有关钽酸盐的研究要少得多。

6.3.1　合成

自首次报道六钽氧簇以来，又有多篇文献探讨六钽氧簇碱金属盐的合成[60-62]，钾盐的合成采用下述方法：按 Ta_2O_5 和 KOH 的质量比 2∶8.5，称取药品，将 Ta_2O_5 放入坩埚底部，KOH 放在上部。将坩埚放入马弗炉中，在 450℃ 左右烧 40min，然后调温至 560℃，继续烧 1h。随后取出坩埚，稍冷后加入少量的蒸馏水浸泡溶解移至大烧杯。向烧杯中加入适量的蒸馏水，加热，溶解。待溶液澄清后趁热抽滤（G3 布氏漏斗），将滤液放入烧瓶中，用旋转蒸发仪蒸发浓缩。然后将浓缩的溶液放入小烧杯中，在冰箱冷冻 12h。最后将 EtOH 和

H_2O 体积比为 1∶1 的溶液 60mL 倒入上述步骤解冻后的样品中,搅拌均匀,抽滤。将抽滤所得的固体加入无水 EtOH 搅拌后再次抽滤 3h,即可得到六钽氧簇的钾盐。

6.3.2 相关化合物

钽元素的惰性不仅表现在其本身难以反应生成新化合物,还表现在它所形成的化合物参与再反应的活性也差。尽管六钽氧簇早在 20 世纪 50 年代就已发现,但以六钽氧簇为原料制得的化合物至今依然很少。2011 年,胡长文课题组报道以六钽氧簇化合物为原料,通过扩散的方法,合成了两例六钽氧簇阴离子参与构成的化合物[63],其分子组成为 $[Cu(en)_2]_4[Ta_6O_{19}] \cdot 14H_2O$ 和 $\{[Cu(1,3\text{-}dap)_2]_2[Cu(1,3\text{-}dap)(H_2O)]_2[Ta_6O_{19}]\} \cdot 8H_2O$,在 $\{[Cu(1,3\text{-}dap)_2]_2[Cu(1,3\text{-}dap)(H_2O)]_2[Ta_6O_{19}]\}$ 中,六钽氧簇阴离子通过表面桥氧原子与铜离子配位,形成六钽氧簇参与配位的化合物(图 6-26)。

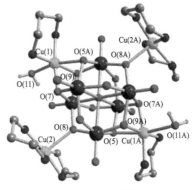

图 6-26 $\{[Cu(1,3\text{-}dap)_2]_2[Cu(1,3\text{-}dap)(H_2O)]_2[Ta_6O_{19}]\}$ 结构

6.4 六钼氧簇

六钼氧簇阴离子由 Fuchs 和 Jahr 在 1968 年通过向酸化的钼酸盐溶液中加入四丁基铵盐首次得到[64],而后又发展了在二甲基甲酰胺中制备六钼氧簇化合物更直接的方法[65],随后六钼氧簇的铵盐、钠盐和铯盐也被报道[66]。

6.4.1 合成

2.5g 钼酸钠溶于 10mL 水中,用 2.9mL 6mol/L 的盐酸酸化,常温下在烧瓶里剧烈搅拌 1min,剧烈搅拌下加入 1.21g 四丁基溴化铵溶于 2mL 水中的溶液,立即生成大量的白色沉淀。反应式为[67]:

$$6Na_2MoO_4 + 10HCl + 2(n\text{-}C_4H_9)_4NBr \longrightarrow [(n\text{-}C_4H_9)_4N]_2[Mo_6O_{19}] + 10NaCl + 2NaBr + 5H_2O$$

得到的白色浆状物加热到 75~80℃,搅拌 45min,搅拌进程中白色固体逐步变为黄色,粗产品用中孔砂芯漏斗抽滤,用水洗涤,空气中干燥的粗产品(2.17g)溶于 80mL 热丙酮中(60℃)并冷却溶液至-20℃,24h 后可以得到黄色晶状产品,抽滤,用乙醚洗涤,真空干燥。

6.4.2 结构

1969 年，Kepert 提出六金属氧簇倾向于形成六个金属氧八面体共边堆积的结构，其阴离子组成应如已知的六铌氧簇和六钽氧簇，具有 M_6O_{19} 组成[68]。1973 年，Allcock 通过测定化合物 $[HN_3P_3(NMe)_6]_2[Mo_6O_{19}]$ 结构，首次从实验上确定了六钼氧簇阴离子 $[Mo_6O_{19}]^{2-}$ 的结构[65]，该阴离子与已知的六铌氧簇阴离子和六钽氧簇阴离子同构，为 Lindqvist 结构，具有 O_h 对称性，六个 MoO_6 八面体通过共边紧密堆积，整个阴离子钼氧键长范围为：端氧 $Mo—O_t$ 1.676～1.678Å；桥氧 $Mo—O_b$ 1.855～2.005Å；中心氧 $Mo—O_c$ 2.312～2.324Å。在后来的研究中，报道了大量六钼氧簇阴离子与不同阳离子结合形成新化合物的晶体数据，由于受阳离子的影响和测试条件的差别，阴离子中的 Mo—O 键长存在一些差别，表 6-4 列出了六钼氧簇阴离子在不同化合物中的 Mo—O 键长。

表 6-4 六钼氧簇阴离子在不同化合物中的 **Mo—O 键长**   单位：Å

化合物	$Mo—O_t$	$Mo—O_b$	$Mo—O_c$	文献
$[HN_3P(NMe_2)_6]_2[Mo_6O_{19}]$	1.677	1.928	2.319	[65]
$[(n\text{-}C_4H_9)_4N]_2[Mo_6O_{19}]$	1.677	1.925	2.319	[69]
$[PPh_3CH_2COOEt]_2[Mo_6O_{19}]$	1.676	1.931	2.321	[70]
$[PPh_3CH_2Ph]_2[Mo_6O_{19}]$	1.676	1.921	2.312	[70]
$[Mo(CO)_2(dpmSe)_2Cl]_2[Mo_6O_{19}] \cdot 4CH_3NO_2$ (dpmSe=$Ph_2PCH_2P(Se)Ph_2$)	1.663	1.936	2.311	[71]
$[TTF]_3[Mo_6O_{19}]$（TTF=四硫富瓦烯）	1.677	1.925	2.327	[72]
$(BEDT\text{-}TTF)_2Mo_6O_{19}$（BEDT-TTF=二硫化乙烯四硫富瓦烯）	1.691	1.928	2.322	[73]
$[(t\text{-}C_4H_9)_4N]_2[Mo_6O_{19}]$	1.673	1.925	2.319	[74]
$[DMDPh\text{-}TTF]_2[Mo_6O_{19}]$（DMDPh-TTF=3,3'-二甲基-4,4'-二苯基-2,2',5,5'-四硫富瓦烯）	1.684	1.925	2.316	[75]
$[TTF]_2[Mo_6O_{19}]$	1.680	1.920	2.316	[76]
$[TTF]_3[Mo_6O_{19}]$	1.686	1.939	2.322	[76]
$[TTP]_2[Mo_6O_{19}]$（TTP=三苯基吡喃鎓）	1.673	1.917	2.3123	[77]
$[Ph_3P=N=PPh_3]_2[Mo_6O_{19}]$	1.672	1.924	2.316	[78]
$[2\text{-}DAPP]_2[Mo_6O_{19}]$（2-DAPP=2-(4-二甲基氨苯基)乙基吡啶鎓）	1.674	1.926	2.319	[79]
$[DAPQ]_2[Mo_6O_{19}]$（DAPQ=4-(4-二甲基氨苯基)乙基喹啉鎓）	1.670	1.923	2.313	[79]

续表

化合物	Mo—O_t	Mo—O_b	Mo—O_c	文献
[Ni(phen)$_3$][Mo$_6$O$_{19}$](phen=1,10-邻菲啰啉)	1.683	1.923	2.316	[80]
[(CH$_3$)$_4$N]$_2$[Mo$_6$O$_{19}$]	1.654	1.916	2.302	[81]
[(CH$_3$)$_4$N]$_2$[Mo$_6$O$_{19}$]	1.679	1.921	2.304	[82]
[Y$_2$(DNBA)$_2$(OAc)$_2$(DMF)$_6$][Mo$_6$O$_{19}$](DNBA=3,5-二硝基苯甲酸)	1.691	1.934	2.3202	[83]
[Na(DB$_{18}$C$_6$)(CH$_3$CN)]$_2$[Mo$_6$O$_{19}$]	1.681	1.927	2.320	[84]

由表 6-4 可知，在已知结构的六钼氧簇形成的化合物中，存在有机铵阳离子、有机给体、四氢富瓦烯类阴离子自由基和配阳离子，下面分别予以介绍。

在四甲基铵阳离子和六钼氧簇阴离子形成的化合物 [(CH$_3$)$_4$N]$_2$[Mo$_6$O$_{19}$]（图 6-27）中[82]，阴、阳离子通过静电引力结合。

由于六钼氧簇阴离子是良好的电子受体，它可以和众多的有机电子给体形成电荷转移盐。在 Kochi 等人报道的由二茂铁阳离子作给体与六钼氧簇形成的电荷转移化合物（图 6-28）中[85]，环戊二烯基上的氢原子与阴离子上的氧原子存在最短接触，最短距离在 2.66～2.76Å，并且漫反射电子光谱图上在 λ_{max}=550nm 处出现了新的荷移带，表明发生了电荷转移。

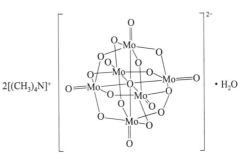

图 6-27　[(CH$_3$)$_4$N]$_2$[Mo$_6$O$_{19}$]结构

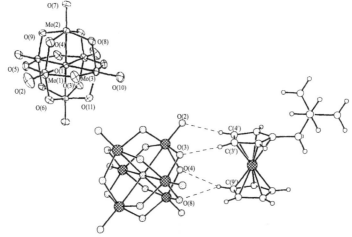

图 6-28　[CpFeCpCH$_2$NMe$_3$]$_2$Mo$_6$O$_{19}$ 结构

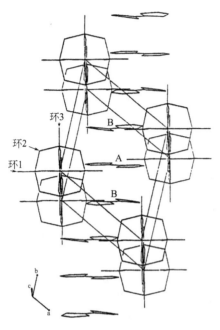

图 6-29　3∶1 型电荷转移盐 [TTF]$_3$[Mo$_6$O$_{19}$] 堆积

六钼氧簇阴离子可作为电子受体与作为电子给体的四氢富瓦烯类分子相结合形成给体-受体的电荷转移化合物，其化学计量比存在 2∶1 和 3∶1 两种类型。Bellitto 研究组在 1991 年首次确定了六钼氧簇阴离子与四氢富瓦烯形成的电荷转移化合物的结构[76]。在 [TTF]$_3$[Mo$_6$O$_{19}$] 的结构中，含有两种取向的 TTF 分子 A 和 B，A 和 B 通过三聚体内 S-S 强接触堆积形成三聚体链…BAB…BAB…，填充在由六钼氧簇阴离子形成的孔道中（图 6-29）。在给体-受体化学计量比为 3∶1 的电荷转移盐中，有机给体部分应该为混合价。通过对 BAB 三聚体进行扩展 Hückel 分子计算可知，两个正电荷分布在三个 TTF 分子上，处于三聚体外缘的两个 B 分子各带 0.5 个正电荷，处于中间的 A 分子带 1 个正电荷。通过改变实验条件，可以得到给体-受体比例为 2∶1 的电荷转移盐。如果在电结晶过程中用六钼氧簇的正四丁基铵盐取代四乙基铵盐，可以得到给体-受体比例为 2∶1 的 [TTF]$_2$[Mo$_6$O$_{19}$]。给体-受体化学计量比为 2∶1 的电荷转移盐的结构不呈现在 3∶1 中的链状结构，而是以阳离子自由基二聚体的形式，分布在六钼氧簇阴离子之间（图 6-30）。

由六钼氧簇阴离子与配位阳离子结合形成的化合物已有多例报道[71,80,83,84,86-88]。在不同的化合物中，六钼氧簇阴离子和配阳离子的结合方式也不尽相同。在与二苯并-18-冠-6-合钠配离子形成的化合物 [Na(DB$_{18}$C$_6$)(CH$_3$CN)]$_2$[Mo$_6$O$_{19}$] 中，六钼氧簇阴离子通过端基氧与配阳离子中的钠相连，形成如图 6-31 所示的担载结构[84]，而在化合物 [Na(DB$_{18}$C$_6$)(H$_2$O)$_{1.5}$]$_2$[Mo$_6$O$_{19}$]·CH$_3$CN 中，六钼氧簇阴离子和配阳离子之间通过静电引力相结合，形成如图 6-32 所示的分立结构[87]。

6.4.3　振动光谱

Rocchiccioli-Deltcheff 等人通过全同位素取代和内坐标分析，研究了 [Mo$_6$O$_{19}$]$^{2-}$ 的振动光谱[89]。图 6-33 为 [Mo$_6$O$_{19}$]$^{2-}$ 的内坐标，具有 162 个内

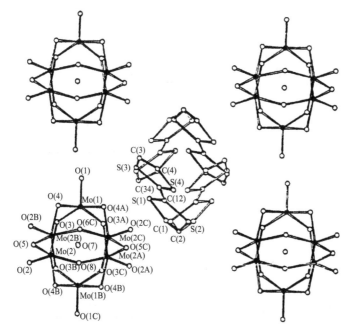

图 6-30　2∶1 型电荷转移盐 [TTF]$_2$[Mo$_6$O$_{19}$] 堆积

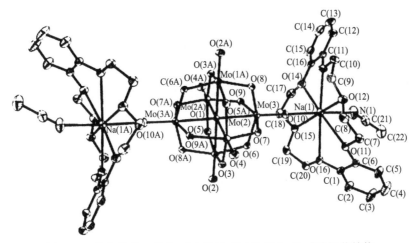

图 6-31　化合物 [Na(DB$_{18}$C$_6$)(CH$_3$CN)]$_2$[Mo$_6$O$_{19}$] 的担载结构

坐标，分别定义为：

$$6 \text{ 个 Mo—O}_c \text{ 键} \quad c$$
$$6 \text{ 个 Mo—O}_t \text{ 键} \quad t$$
$$24 \text{ 个 Mo—O}_b \text{ 键} \quad b$$
$$12 \text{ 个 MoO}_c\text{Mo 角} \quad \alpha$$
$$24 \text{ 个 O}_c\text{MoO}_b \text{ 角} \quad \beta$$

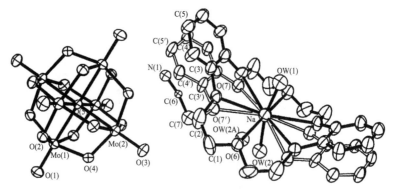

图 6-32 化合物 $[Na(DB_{18}C_6)(H_2O)_{1.5}]_2[Mo_6O_{19}] \cdot CH_3CN$ 的分立结构

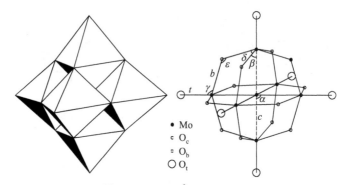

图 6-33 $Mo_6O_{19}^{2-}$ 的内坐标

24 个 O_bMoO_t 角 γ

24 个 O_bMoO_b 角 δ

12 个 MoO_bMo 角 ε

6 个扭转坐标(以 $Mo-O_c$ 为共线的二面角)

24 个扭转坐标(以 $Mo-O_b$ 为共线的二面角)

按照振动模式分析[89,90]，$[Mo_6O_{19}]^{2-}$ 应可观察到 7 个红外振动吸收峰。$[TBA]_2[Mo_6O_{19}]$ 的红外光谱表明（图 6-34），只有六个 $[Mo_6O_{19}]^{2-}$ 的振动吸收峰可观察到，未观察到的吸收峰可能存在于 200cm^{-1} 以下。通过内坐标分析和同位素取代研究可以指认 956 cm^{-1} 的吸收峰基本为纯的 $Mo-O_t$ 键的伸缩振动；796cm^{-1} 的吸收峰主要为 $Mo-O_b$ 键的伸缩振动，但其中也有 t 伸缩振动和 ν 弯曲振动的贡献；598 cm^{-1} 处的吸收峰主要为 ν 振动，其中有 $Mo-O_b$ 键振动的贡献；432cm^{-1} 处的吸收峰组成复杂，主要包括了 ν、b、ε 和 δ 振动的贡献；352 cm^{-1} 处的振动峰几乎为纯的 $Mo-O_c$ 键的振动吸收；217cm^{-1} 的振动峰主要为 δ 弯曲振动的贡献。

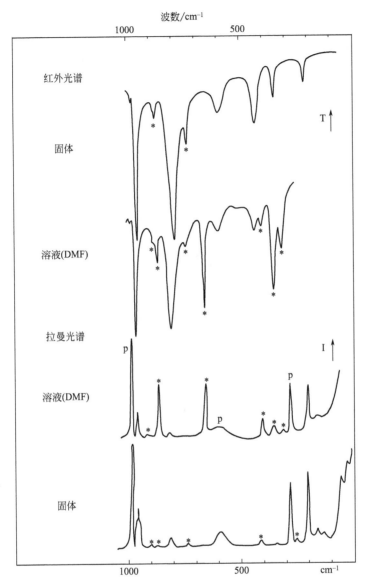

图 6-34 化合物 [TBA]$_2$[Mo$_6$O$_{19}$] 红外及拉曼光谱

粗略看来，956 cm^{-1} 的吸收峰对应于 Mo—O$_t$ 键的振动，796 cm^{-1} 的吸收峰对应于 Mo—O$_b$ 键的振动，352 cm^{-1} 的吸收峰对应于 Mo—O$_c$ 键的振动。

6.4.4 衍生物

六钼氧簇阴离子上的端氧具有较强的反应活性，可与包括氮烯基、亚肼基、偶氮羟基、亚氨基、亚硝基和金属有机基团在内的基团形成衍生物。Peng 和 Proust 等都对该领域的工作进行过总结[91,92]，下面分别进行介绍。

(1) 氮烯衍生物

1986 年，Zubieta 报道了由 $[Bu_4N]_2[Mo_6O_{19}]$ 在无水苯溶剂中与等量的苯肼和三乙胺反应回流 3h，得到深红色溶液，缓慢蒸发得到黑红色晶体。元素分析表明该晶体分子组成为 $[Bu_4N]_3[Mo_6O_{18}(NNC_6H_5)]^{[93]}$。该化合物为首例六钼氧簇端基氧被含氮有机基团取代的衍生物[80]。红外光谱表明化合物中 N—N 振动峰 $\nu_{(N-N)}=1588cm^{-1}$，端氧振动吸收峰 $\nu_{Mo-O_t}=935cm^{-1}$，桥氧振动吸收峰 $\nu_{Mo-O_b}=790\ cm^{-1}$。晶体结构分析表明，氮烯基中的氮原子取代端基氧原子的位置，通过钼氮键与六钼氧簇骨架相连，结构如图 6-35 所示。

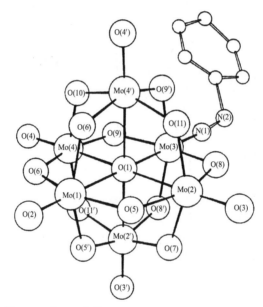

图 6-35　$[Bu_4N]_3[Mo_6O_{18}(NNC_6H_5)]$ 原子连接图

Bank 等人通过五氟代苯肼（$C_6F_3HN—NH_2$）与八钼酸四丁基铵 $[Bu_4N]_4[Mo_8O_{26}]$ 反应，制得了分子组成为 $[(n-C_4H_9)_4N]_3[Mo_6O_{18}(NNC_6F_5)]^{[94]}$ 的化合物，Gouzerh 及其合作者以 $[Bu_4N]_4[Mo_8O_{26}]$ 和芳基肼衍生物为原料，制得了一系列分子组成为 $[n-Bu_4N]_3[Mo_6O_{18}(N_2Ar)]$ 的衍生物[95]，其中 Ar= C_5H_5[93]、C_6F_5[94]、$C_6H_4-p-NO_2$[96] 和 $C_6H_3-o,p-(NO_2)_2$[97] 已被分别报道。

结构分析表明，$[Mo_6O_{18}(N_2Ar)]^{3-}$ 中的 $[Mo(N_2Ar)]^{3+}$ 和 $[Mo_6O_{18}(NO)]^{3-}$ 中的 $[Mo(NO)]^{3+}$ 基团的成键方式相同，在这种配位方式中氮烯基和亚硝基配体通常分别被看作 RN_2^+ 和 NO^+ 基团，因此 $[Mo(N_2Ar)]^{3+}$ 基团中应该含有 Mo^{II} 中心。$[Mo_6O_{18}(N_2Ar)]^{3-}$ 中的氮烯基配体在配合物中表现出 Mo—N—N 的线性排列和较短的 Mo—N、N—N 键长（表 6-5），表明多重键的存在。

β 氮原子的键角在 120°左右表明该原子为 sp² 杂化。

表 6-5 $[n\text{-}Bu_4N]_3[Mo_6O_{18}(N_2Ar)]$ 中 $[Mo(N_2Ar)]$ 的结构数据

化合物	Mo—N/Å	N—N/Å	Mo—N—N/(°)	文献
$(n\text{-}Bu_4N)_3[Mo_6O_{18}(N_2C_6H_5)]$	1.76(2)	1.31(3)	178.7(19)	[93]
$(n\text{-}Bu_4N)_3[Mo_6O_{18}(N_2C_6F_5)]$	1.75(2)	1.30(3)	170.6(21)	[94]
$(n\text{-}Bu_4N)_3[Mo_6O_{18}(N_2C_6H_4\text{-}o\text{-}NO_2)]$	1.75(1)	1.285(14)	173.9(10)	[95]
$(n\text{-}Bu_4N)_3[Mo_6O_{18}(N_2C_6H_4\text{-}p\text{-}CO_2H)]$	1.76(1)	1.246(13)	174.7(10)	[95]
$(n\text{-}Bu_4N)_3[Mo_6O_{18}(N_2C_6H_4\text{-}p\text{-}NO_2)]$	1.74(2)	1.29(2)	178(2)	[96]
$(n\text{-}Bu_4N)_3[Mo_6O_{18}\{N_2C_6H_3\text{-}o,p\text{-}(NO_2)_2\}]$	1.762	1.292	176.03	[97]

(2) 亚肼衍生物

由 $[MoCl_4(NNMePh)]$ 与 $[Bu_4^nN]_2[Mo_2O_7]$ 在甲醇中反应,可以得到黑色晶体 $(Bu_4^nN)_2[Mo_6O_{18}(NNMePh)]$[98],其反应如下:

$$2[MoCl_4(NNMePh)] + 5[Mo_2O_7]^{2-} + H_2O \longrightarrow 2[Mo_6O_{18}(NNMePh)]^{2-} + 6Cl^- + 2HCl$$

另外,由 $[Mo_6O_{18}(NNMePh)]^{3-}$ 与碘甲烷在 CH_2Cl_2 中反应,也可以得到 $[Mo_6O_{18}(NNMePh)]^{2-}$。

化合物的红外光谱在 1589cm^{-1} 处出现了 $\nu_{(N=N)}$ 的振动吸收峰,在 943cm^{-1} 和 780cm^{-1} 处分别出现了端氧 $\nu_{(Mo-O_t)}$ 和桥氧 $\nu_{(Mo-O_b)}$ 的振动吸收峰。

六钼氧簇亚肼衍生阴离子结构如图 6-36 所示,六钼氧簇阴离子中的一个端基氧原子被亚肼基 $(-NNMePh)^{2-}$ 取代。

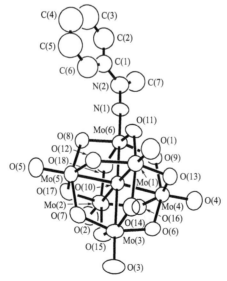

图 6-36 化合物中 $(Bu_4^nN)_2[Mo_6O_{18}(NNMePh)]$ 阴离子结构

(3) 重氮烷衍生物

$[Bu_4N]_2[Mo_6O_{19}]$ 与三苯磷嗪在80℃的吡啶溶剂中反应5h生成 $Ph_3P=O$ 和六钼氧簇的偶氮烷衍生物[99]，反应式如下：

$$[Bu_4N]_2[Mo_6O_{19}] + Ph_3P=NN=C(C_6H_4OCH_3)CH_3$$
$$\xrightarrow[5h]{\text{吡啶},80℃} [Bu_4N]_2[Mo_6O_{18}(NN-\underset{CH_3}{C}-C_6H_4-OCH_3)] + Ph_3P=O$$

蒸发掉溶剂，用苯洗去副产物，可得到深橙色粉末产物，用乙醚向乙腈溶液中扩散的方法，可以得到适于结构测定的单晶。衍生物阴离子结构如图6-37所示，偶氮烷配体取代了六钼氧簇阴离子的一个端基氧的位置，以单齿方式与六钼氧簇骨架相连。

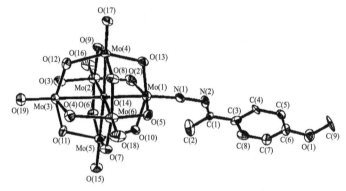

图6-37　$[Mo_6O_{18}(N_2CMeC_6H_4OMe)]^{2-}$ 结构

(4) 亚氨基衍生物

1992年，Maatta及其合作者首次报道了六钼氧簇亚氨基衍生物的制备[100]，随后有大量文献报道了该领域工作，到目前为止，六钼氧簇亚胺类衍生物是其所有衍生物中研究得最为广泛的一类[101-124]。

过去一段时间一直认为亚氨基衍生物的生成，主要基于下述三种类型的反应。

① 六钼氧簇阴离子与膦亚胺反应：

$$[Mo_6O_{19}]^{2-} + Ph_3P=NAr \xrightarrow[85℃,48h]{\text{吡啶}} [Mo_6O_{18}(NAr)]^{2-} + Ph_3P=O$$

② 六钼氧簇阴离子与异氰酸酯反应：

$$[Mo_6O_{19}]^{2-} + xArNCO \xrightarrow[110℃,2\sim13d]{\text{吡啶}} [Mo_6O_{19-x}(NAr)_x]^{2-} + xCO_2$$

③ 六钼氧簇阴离子与芳胺反应：

$$[Mo_6O_{19}]^{2-} + ArNH_2 \xrightarrow[150℃]{PhCN, Et_3N} [Mo_6O_{18}(NAr)]^{2-} + H_2O$$

最近的研究表明，从 α 结构八钼氧簇四丁基铵盐出发，通过与芳胺衍生物反应，也可以制得六钼氧簇的亚氨基衍生物[101-104]。

通过调整六钼氧簇四丁基铵盐和 2,6-二异丙基苯基异氰酸酯的化学计量比，Maatta 及其合作者制得了一系列六钼氧簇阴离子中的端基氧被亚氨基逐个取代的衍生物[106,108,110]。

$$[Bu_4N]_2[Mo_6O_{19}] + xArNCO \xrightarrow[\triangle]{Py} [Bu_4N]_2[Mo_6O_{19-x}(NAr)_x] + xCO_2$$

(NAr=2,6-二异丙基苯亚氨基配体,x=1~6)

图 6-38 给出了六钼氧簇阴离子中的端基氧从单取代到六取代的阴离子结构。

六钼氧簇亚氨基衍生物的电化学性质研究表明[110]，对六钼氧簇单取代亚氨基衍生物来讲，其还原电位随氮原子上有机基团给电子能力增强而变大。对 $[Mo_6O_{19-x}(NAr)_x]^{2-}$ 系列化合物来讲，随着亚氨基取代个数的增加，六钼氧簇衍生物阴离子表面的电荷密度将会升高，其还原电位会变得更低。图 6-39 给出了六钼氧簇衍生物阴离子随取代基个数的增加还原电位的变化情况。从图中可以看出，六钼氧簇阴离子中每增加一个 NAr 取代基，其还原电位降低约 220mV，当 NAr 取代基增加到 5 个时，还原变得不可逆。

(5) 亚硝基衍生物

将 $[Bu_4N]_2[Mo_4O_{12}\{NC—CN=CH—C(NH_2)NO\}_2]$ 和羟胺在甲醇中回流，冷却溶液后可以得到紫色晶体，晶体结构分析表明，紫色晶体分子组成为 $(Bu_4N)_2[Mo_5Na(NO)O_{13}(OCH_3)_4] \cdot 4CH_3OH$。将紫色晶体溶于二氯甲烷或乙腈中，即可自发地转化为分子组成为 $[Bu_4N]_3[Mo_6O_{18}(NO)]$ 的化合物[125]。由 $[Bu_4N]_3[Mo_6O_{18}(NO)]$ 与硫酸二甲酯在乙腈中回流，可以得到分子组成为 $[Bu_4N]_2[Mo_6O_{17}(OMe)(NO)]$ 的衍生物[126]。由于该类化合物分子在固体中存在大量无序结构，至今没有得到该类化合物的晶体结构数据。

(6) 金属有机衍生物

$[(\eta-C_5Me_5)Mo(CO)_2]_2$ 在 $CHCl_3$ 中用空气氧化，可以得到分子组成为 $[C_5Me_5O][(\eta-C_5Me_5)Mo_6O_{18}]$ 的化合物[127]，该化合物阴离子结构如图 6-40 所示。钼氧簇阴离子 $[Mo_6O_{19}]^{2-}$ 中的一个端基氧原子被 C_5Me_5 基团取代，形成六钼氧簇的 C_5Me_5 衍生物。由 $[Bu_4N][MoC_5Me_5O_3]$ 与 $[Bu_4N]_2[Mo_4O_{13}(OMe)_4Cl_2]$ 在甲醇中回流，也可以得到含有六钼氧簇阴离子金属有机衍生物的化合物 $[Bu_4N][(\eta-C_5Me_5)Mo_6O_{18}]$[128]。

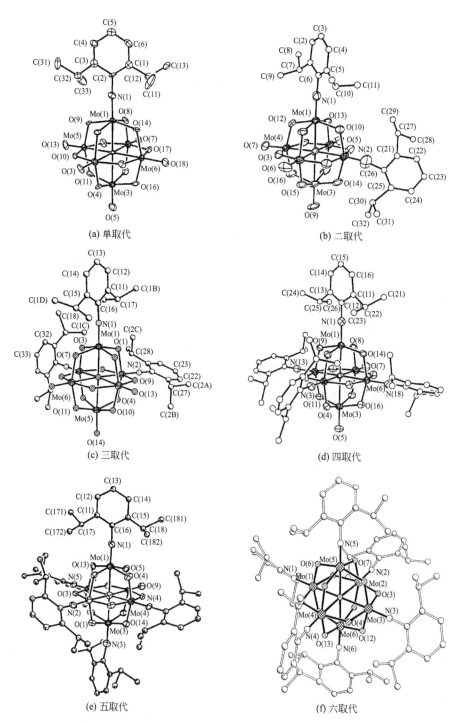

图 6-38 六钼氧簇阴离子 2,6-二异丙基苯亚氨基衍生物结构

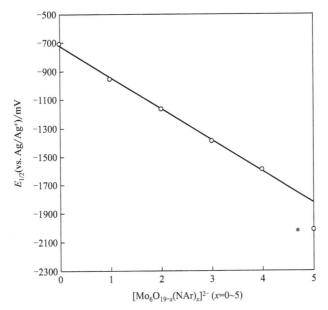

图 6-39　$[Mo_6O_{19-x}(NAr)_x]^{2-}$ 氧化还原性随取代基个数（x）变化情况

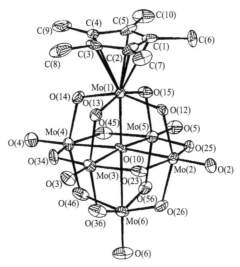

图 6-40　化合物 $[Bu_4N][(\eta\text{-}C_5Me_5)Mo_6O_{18}]$ 中阴离子结构

6.5　六钨氧簇

六钨氧簇与六铌氧簇、六钽氧簇及六钼氧簇同构。1968年，Fuchs通过钨酸酯的控制水解，首次制得了六钨氧簇的四丁基铵盐[64]。

6.5.1 合成

六钨氧簇化合物的合成已有多个报道[64,129,130]，通常四丁基铵盐可采用下述方法制备：33g（100mmol）钨酸钠，40mL乙酸酐和30mL N,N-二甲基甲酰胺在250mL的烧瓶中100℃搅拌3h，搅拌下加入20mL乙酸酐和18mL12mol/L的盐酸溶于50mLDMF的溶液，得到的混合物用中速滤纸过滤除去白色固体，用50mL甲醇洗涤固体，将澄清的滤液冷却至室温，快速搅拌下加入15g（47mmol）四丁基溴化铵溶于50mL甲醇中的溶液，得到白色沉淀，悬浊液搅拌5min，抽滤分离出产品，用20mL甲醇和50mL乙醚洗涤。重结晶可以在二甲基亚砜溶剂中进行。

6.5.2 谱学表征

$[(n\text{-}C_4H_9)_4N]_2W_6O_{19}$ 红外光谱如图6-41所示，其中位于975cm^{-1}、812cm^{-1}、588cm^{-1} 和445cm^{-1} 的吸收峰为 $[W_6O_{19}]^{2-}$ 特征振动吸收峰。$[W_6O_{19}]^{2-}$ 的UV-Vis光谱如图6-42所示，位于278nm的吸收峰为该阴离子的特征吸收峰[131]。

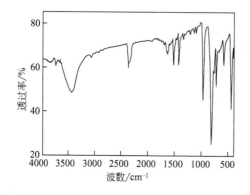

图6-41 $[(n\text{-}C_4H_9)_4N]_2W_6O_{19}$ 红外光谱

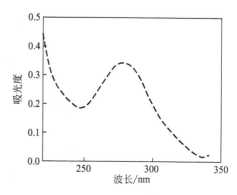

图6-42 $[W_6O_{19}]^{2-}$ 的UV-Vis光谱

6.5.3 结构研究

1978年，Fuchs分离出了由 $[W_6O_{19}]^{2-}$ 形成的盐并测定了化合物的晶体结构[132]。该化合物阴离子与六铌氧簇、六钽氧簇和六钼氧簇同构，具有Lindqvist结构（图6-2）。在后来的研究工作中，报道了多例由六钨氧簇阴离子 $[W_6O_{19}]^{2-}$ 参与形成的化合物的晶体结构[133-143]，数据列于表6-6。

表 6-6　$[W_6O_{19}]^{2-}$ 在不同化合物中的键长　　　　　　单位：Å

化合物	W—O_t	W—O_b	W—O_c	文献
$[(t\text{-}C_4H_9NC)_7W][W_6O_{19}]$	1.693	1.907	2.315	[133]
$[TMTTF]W_6O_{19}$	1.691	1.926	2.327	[134]
$[n\text{-}Bu_4N]_4[Ag_2I_4][W_6O_{19}]$	1.733	1.911	2.316	[135]
$[(Ph)_4P]_2[W_6O_{19}]$	1.714	1.917	2.323	[136]
$[WI(CO)(L\text{-}P,P',P'')(\eta^2\text{-}MeC_2Me)]_2[W_6O_{19}]$	1.680	1.943	2.352	[137]
$[(CH_3Py)_2H]_2W_6O_{19}$	1.697	1.920	2.324	[138]
$[Cu(phen)_3][W_6O_{19}]$	1.700	1.920	2.324	[139]
$[La_2(DNBA)_4(DMF)_8][W_6O_{19}]$	1.698	1.921	2.325	[140]
$(PPN)_2[W_6O_{19}]$	1.690	1.920	2.320	[141]

由表 6-4 和表 6-6 可知，目前报道的由六钨氧簇形成的化合物比六钼氧簇形成的化合物的晶体结构数据要少。在六钨氧簇形成的化合物中，阳离子有金属有机阳离子、自由基阳离子、有机铵阴离子及金属配离子等。在这些化合物中，阴离子与阳离子之间通过静电引力相结合，阴离子结构保持不变。

2006 年，Kortz 及其合作者报道了一个组成为 $[((CH_3)_2Sn)_2(W_6O_{22})]^{4-}$ 的阴离子[142]，其中含有六钨氧簇基团 $[W_6O_{22}]^{8-}$，结构如图 6-43 所示。在该阴离子中，三个八面体通过共边连成一组三金属簇，两组三金属簇通过两个八面体的三条棱共边相连。该阴离子与 $[\{Cu(2,2'\text{-bpy})\}_6(Mo_6O_{22})][GeMo_{12}O_{40}]$ 中所含的六钼氧簇的结构相似[143]。由于该类结构的阴离子中包含具有两个以上端氧的八面体，目前所见到的都以配体的形式存在，是否存在独立结构的六钨氧簇 $[W_6O_{22}]^{8-}$，还有待进一步研究。

6.5.4　衍生物

与六钼氧簇相似，六钨氧簇阴离子也可以形成骨架上的端基氧或骨架上的金属原子被其他基团取代的衍生物，但由于钨原子本身的特性，六钨氧簇形成的衍生物比六钼氧簇形成的衍生物要少。

(1) 亚胺衍生物

六钼氧簇亚胺衍生物可由六钼氧簇阴离子直接反应生成。由于六钨氧簇阴离子中端基氧的稳定性，到目前为止，还未见由六钨氧簇阴离子直接反应生成六钨氧簇亚氨基衍生物的报道，目前报道的唯一六钨氧簇亚氨基衍生物阴离子是由 $[Bu_4N]_2WO_4$ 和 ArNCO 在二氯乙烷或吡啶中制得的[144]：

$$6[Bu_4N]_2[WO_4]+ArNCO+5H_2O \longrightarrow [Bu_4N]_2[W_6O_{18}(NAr)]+CO_2+10[Bu_4N][OH]$$

通过该方法，Maatta 及其合作者制得了结构如图 6-44 所示的六钨氧簇衍生物阴离子。

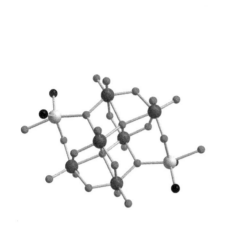

图 6-43 六钨氧簇阴离子 $[W_6O_{22}]^{8-}$ 结构　　图 6-44 $[W_6O_{18}(NAr)]^{2-}$ 结构

(2) 重氮烷衍生物

与亚胺衍生物一样，重氮烷衍生物也难以由六钨氧簇阴离子直接制得。由 $[Bu_4]_2[WO_4]$ 和 $Ph_3P=NN=C(Me)C_6H_4OMe$ 在二氯乙烷中回流反应 3d，生成橙红色溶液，经过一系列处理，可以得到分子组成为 $[Bu_4N]_2[W_6O_{18}(N_2C(Me)C_6H_4OMe)]$ 的晶体，产率大约为 10%[145]。阴离子结构如图 6-45 所示。

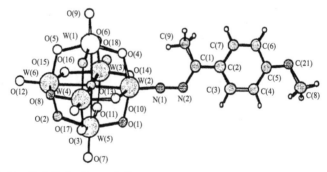

图 6-45 化合物中 $[Bu_4N]_2[W_6O_{18}(N_2C(Me)C_6H_4OMe)]$ 阴离子结构

(3) 亚硝基衍生物

由 $(n\text{-}Bu_4N)_2WO_4$ 和 $[W(NO)Cl_3(MeCN)_2]$ 在重蒸乙腈中反应，可以得到分子组成为 $[n\text{-}Bu_4N]_2[W_5O_{18}(NO)]$ 的深绿色晶体。由 $(n\text{-}Bu_4N)_2WO_4$ 和 $[Mo(NO)\{MeC(NH_2)NO\}(acac)_2]$ 在乙腈中反应，可以得到分子组成为 $(n\text{-}Bu_4N)_3[W_5O_{18}\{Mo(NO)\}]$ 的浅绿色晶体[146]。该类化合物的阴离子应具

(4) 金属有机衍生物

1985 年，Klemperer 及其合作者报道了组成为 $[(n\text{-}C_4H_9)_4N]_3[(\eta^5\text{-}C_5H_5)Ti(W_5O_{18})]$ 的化合物[147]。该化合物由 $[(n\text{-}C_4H_9)_4N]_2WO_4$ 和二氯化二茂钛在乙腈中反应生成，反应方程式如下：

$$5WO_4^{2-} + 5H^+ + (C_5H_5)_2TiCl_2 \longrightarrow [(C_5H_5)Ti(W_5O_{18})]^{3-} + C_5H_6 + 2Cl^- + 2H_2O$$

结构分析表明，阴离子具有图 6-47 所示的结构，在该阴离子中，Lindqvist 结构中的一个钨原子被钛原子取代，钛原子端基氧的位置由环戊二烯基占据。

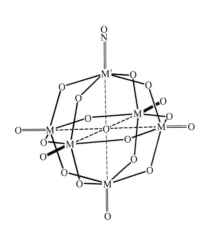

图 6-46 $(n\text{-}Bu_4N)_3$ $[W_5O_{18}\{Mo(NO)\}]$ 结构

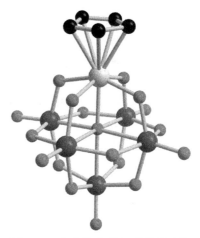

图 6-47 化合物中 $[(n\text{-}C_4H_9)_4N]_3$ $[(\eta^5\text{-}C_5H_5)Ti(W_5O_{18})]$ 阴离子结构

由 $[Cp^*W(CO)_2]$ 在一定的条件下发生脱羰基反应，可以生成分子组成为 $(C_5Me_5)_2W_6O_{17}$ 的化合物[148]。在该化合物的分子中，两个五甲基环戊二烯分别占据两个钨原子的端基氧位置，形成如图 6-48 所示的结构。

图 6-48 化合物 $(C_5Me_5)_2W_6O_{17}$ 结构

参考文献

[1] Lindqvist I. The structure of the hexaniobate ion in $7Na_2O \cdot 6Nb_2O_5 \cdot 32H_2O$. Arkiv Kemi, 1952, 5: 247.

[2] Britton H T S, Robinson R A. Physicochemical studies of complex acids. Part II. Vanadic acid. J Chem Soc, 1930: 1261-1274.

[3] Brintzinger H, Wallach J. Die in alkalischen lösungen existierenden polyvanadationen. Z Anorg Allge Chem, 1935, 224: 103.

[4] Newman L, Quinlan K P. A spectrophotometric investigation of vanadium (V) species in acidic solutions. J Am Chem Soc, 1959, 81 (3): 547-549.

[5] Wittaker M P, Asay J, Eyring E M. A kinetic study of vanadate polymerization in aqueous solution. J Phys Chem, 1966, 70 (4): 1005.

[6] Naumann W, Hallada C J. A study of the chemistry of the polyvanadates using salt cryoscopy. Inorg Chem, 1964, 3 (1): 70-77.

[7] Howarth W, Richards R E. Nuclear magnetic resonance study of polyvanadate equilibria by use of vanadium-51. J Chem Soc, 1965: 864.

[8] Larson J W. Thermochemistry of vanadium (5^+) in aqueous solutions. J Chem Eng Data, 1995, 40: 1276-1280.

[9] Dullberg P. On the behavior of the vanadates in aqueous solution. Z physic Chem, 1903, 45 (2): 129-181.

[10] Chae H K, Klemperer W G, Day V W. An organometal hydroxide route to $[(C_5Me_5)Rh]_4(V_6O_{19})$. Inorg Chem, 1989, 28: 1423-1424.

[11] Hayashi Y, Ozawa Y, Isobe K. The first "vanadate hexamer" capped by four pentamethylcyclopentadienyl-rhodium or iridium groups. Chem Lett, 1989, 3: 425.

[12] Hayashi Y, Ozawa Y, Isobe K. Site-selective oxygen exchange and substitution of organometallic groups in an amphiphilic quadruple-cubane-type cluster. Synthesis and molecular structure of $[(MCp^*)_4V_6O_{19}]$ (M = rhodium, iridium). Inorg Chem, 1991, 30 (5): 1025-1033.

[13] Chen Qin, Zubieta J. Synthesis and structural characterization of a polyoxovanadate coordination complex with a hexametalate core: $[(n-C_4H_9)_4N]_2[V_6O_{13}\{O_2NC(CH_2O)_3\}_2]$. Inorg Chem, 1990, 29 (8): 1456-1458.

[14] Chen Qin, Zubieta J. Structural investigations of the hexavanadium core $\{V_6O_{19}\}$ in "oxidized", mixed valence and "reduced" clusters of the type $[V_{6-n}^V V_n^{IV} O_{13-n}(OH)_n\{(OCH_2)_3CR\}_2]^{2-}$, $n=0$, 3 and 6. Inorg Chim Acta, 1992, 198-200: 95-110.

[15] Chen Qin, Goshorn D P, Scholes C P, Tan Xiao-ling, Zubieta J. Coordination compounds of polyoxovanadates with a hexametalate core. Chemical and structural characterization of $[V_6^V O_{13}\{(OCH_2)_3CR\}_2]^{2-}$, $[V_6^V O_{11}(OH)_2\{(OCH_2)_3CR\}_2]$, $[V_4^{IV}V_2^V O_9(OH)_4$

$\{(OCH_2)_3CR\}_2]^{2-}$, and $[V_6^{IV}O_7(OH)_6\{(OCH_2)_3CR\}_2]^{2-}$. J Am Chem Soc, 1992, 114: 4667-4681.

[16] Chen Qin, Zubieta J. A novel hexavanadate core: synthesis and structure of the mixed valence cluster $[V_6O_8\{(OCH_2)_3CEt\}_2\{(OCH_2)_2C(CH_2OH)(Et)\}_4]^{2-}$ and a comparison with the hexametallate core of $[V_6O_{13}(OMe)_3\{(OCH_2)_3CCH_2OH\}]^{2-}$. J Chem Soc Chem Commun, 1993: 1180.

[17] Muller A, Meyer J, Bogge H, Stammler A, Botar A. *Cis-/trans*-isomerism of bis-(trisalkoxy)-hexavanadates: *cis*-Na$_2$ $[V_6^{IV}O_7(OH)_6\{(OCH_2)_3CCH_2OH\}_2] \cdot 8H_2O$, *cis*-$[CN_3H_6]_3[V^{IV}V_5^VO_{13}\{(OCH_2)_3CCH_2OH\}] \cdot 4.5H_2O$ and *trans*-$(CN_3H_6)_2[V_6^VO_{13}\{(OCH_2)_3CCH_2OH\}_2] \cdot H_2O$. Z Anorg Allg Chem, 1995, 621: 1818-1831.

[18] Kessler V G, Seisenbaeva G A. The mystery of $VO(OEt)_3$ conversion on microhydrolysis disclosed: the X-ray single crystal study of $V_6O_7(OEt)_{12}$. Inorg Chem Commun, 2003: 203-204.

[19] Gong Yun, Hu Changwen, Li Hui. Synthesis and crystal structure of a novel organic-inorganic hybrid hexavanadate $[(phen)_4V_6O_{12}(CH_3OH)_4] \cdot 2CH_3OH \cdot 4H_2O$. J Mol Struct, 2005, 749: 31-35.

[20] Piepenbrink M, Triller M U, Gorman N H J, Krebs B. Bridging the gap between polyoxometalates and classic coordination compounds: a novel type of hexavanadate complex. Angew Chem Int Ed, 2002, 41 (14): 2523.

[21] Kurata T, Uehara A, Hayashi Y, Isobe K. Cyclic polyvanadates incorporating template transition metal cationic species: synthesis and structures of hexavanadate $[PdV_6O_{18}]^{4-}$, octavanadate $[Cu_2V_8O_{24}]^{4-}$, and decavanadate $[Ni_4V_{10}O_{30}(OH)_2(H_2O)_6]^{4-}$. Inorg Chem, 2005, 44, 2524-2530.

[22] Han J W, Hardcastle K I, Hill C L. Redox-active coordination polymers from esterified hexavanadate units and divalent metal cations. Eur J Inorg Chem, 2006: 2598-2603.

[23] Liu Yan-Cheng, Chen Zhen-Feng, Shi Shao-Ming, Luo Hai-Sheng, Zhong Di-Chang, Zou Hong-Li, Liang Hong. Synthesis, crystal structure of polyoxovanadate complex of ciprofloxacin: $V_4O_{10}(\mu-O)_2[VO(H-Ciprof)_2]_2 \cdot 13H_2O$ by hydrothermal reaction. Inorg Chem Commun, 2007, 10: 1269-1272.

[24] Han Jong Woo, Hill C L. A coordination network that catalyzes O_2-based oxidations. J Am Chem Soc, 2007, 129: 15094-15095.

[25] Zhang Gui Ling, Li Yan Tuan, Luo Xu Qiang, Wu Zhi Yong. Synthesis, structural characterization and luminescent properties of a mixed valence hexavanadate cluster. Z Anorg Allg Chem, 2008, 634: 1161-1165.

[26] Aronica C, Chastanet G, Zueva E, Borshch S A, Clemente-Juan J M, Luneau D A. Mixed-valence polyoxovanadate (Ⅲ, Ⅳ) cluster with a calixarene cap exhibiting ferromagnetic V(Ⅲ)-V(Ⅳ) interactions. J Am Chem Soc, 2008, 130: 2365-2371.

[27] Schulz J, Gyepes R, Cisarova I, Stepnicka P. Synthesis, structural characterisation and

bonding in an anionic hexavanadate bearing redox-active ferrocenyl groups at the periphery. New J Chem, 2010, 34: 2749-2756.

[28] Marino N, Lloret F, Julve M, Doyle R P. Synthetically persistent, self assembled $[V_2^{IV}V_4^{V}]$ polyoxovanadates: facile synthesis, structure and magnetic analysis. Dalton Trans, 2011, 40: 12248.

[29] Wu Pingfan, Xiao Zicheng, Zhang Jin, Hao Jian, Chen Jiake, Yin Panchao, Wei Yongge. DMAP-catalyzed esterification of pentaerythritol-derivatized POMs: a new route for the functionalization of polyoxometalates. Chem Commun, 2011, 47: 5557-5559.

[30] Yin Panchao, Wu Pingfan, Xiao Zicheng, Li Dong, Bitterlich E, Zhang Jin, Cheng Peng, Vezenov D V, Liu Tianbo, Yongge Wei. A double-tailed fluorescent surfactant with a hexavanadate cluster as the head group. Angew Chem Int Ed, 2011, 50: 2521-2525.

[31] Santoni M-P, Pal A K, Hanan G S, Tang M-C, Venne K, Furtos A, Ménard-Tremblay P, Malveau C, Hasenknopf B. Coordination-driven self-assembly of polyoxometalates into discrete supramolecular triangles. Chem Commun, 2012, 48: 200-202.

[32] Champetier M G. Chimie Minérale-Mise en evidence etétude de deux nouveaux niobates doubles de potassium et d'ammonium//Spinner M Bernard, Acad C R. Paris: Serie, 1969, 268 (28): 1870-1873.

[33] Décot-Albert M, Dartiguenave M, Dartiguenave Y. N°519. Isopolyanions des éléments de la colonne V_B. Préparation et étude de I'hexaniobate de tris (tétramminecuivre II). Bulletin De La Société Chimique De France, 1969, 9: 3031-3034.

[34] Muller M. Niobates et tantalates de sodium hydratés: cas exceptionnel d'*isomorphism de masse*. Caractérisation de nouvelles familles de sels à 1' état solide. Revue de Chimie minérale, 1970, 7: 359-411.

[35] Goiffon A, Philippot E, Maurin M. Structure crystalline du niobate 7/6 de sodium $(Na_7)(H_3O)Nb_6O_{19} \cdot 14H_2O$. Revue de Chimie Minérale, 1980, 17: 466-476.

[36] Marty A, Goiffon A, Spinner B. Relations structurales entre les isopolyanions du niobium V. Nouvelles méthodes de synthèse des sels correspondants. Nouvelle voie de synthèse de complexes chélatés du niobium V. Revue de Chimie Minérale, 1977, 14: 347-354.

[37] Spinner B. Relations entre les isopolyanions du niobium V et les peroxoniobates alcalins. Revue de Chimie Minérale, 1969, 6: 319-361.

[38] Goiffon M A, Spinner B. N°455-Édifications structurales des isopolyanions du niobium V et du tungstèn VI. Bulletin De La Société Chimique De France, 1975, 11-12: 2435-2441.

[39] Larbi S S, Bodiot D, Spinner B. Nouveaux isopolyanions du niobium V dans des solutions aqueuses de niobates de tétraméthyl(-éthyl) ammonium. Revue de Chimie Minérale, 1976, 13: 497-507.

[40] Flynn C M, Jr Stucky G D. Sodium 6-niobo (ethylenediamine) cobaltate (III) and its chromate (III) analog. Inorg Chem, 1969, 8 (1): 178-180.

[41] Flynn C M, Jr Stucky G D. Heteropolyniobate complexes of manganese (Ⅳ) and nickel (Ⅳ). Inorg Chem, 1968, 7: 332-334.

[42] Farrell F J, Maroni V A, Spiro T G. Vibrational analysis for $Nb_6O_{19}^{8-}$ and $Ta_6O_{19}^{8-}$ and the Raman intensity criterion for metal-metal interaction. Inorg Chem, 1969, 8 (12): 2638-2642.

[43] Mattes R, Bierbusse H, Fuchs J. Schwingungsspektren und kraftkonstanten von polyanionen mit M_6O_{19}-gruppen. Z Anorg Allg Chem, 1971, 385: 230.

[44] Rocchicciol-Deltcheff C, Touvenot R, Dabbabi M. Etude de la structure des niobotungstates $Nb_nW_{6-n}O_{19}^{-2-n}$ au moyen des spectres de vibration. Spectrochim Acta, 1977, 33A: 143-154.

[45] Alam T M, Nyman M, Cherry B R, Segall J M, Lybarger L E. Multinuclear NMR investigations of the oxygen, water, and hydroxyl environments in sodium hexaniobate. J Am Chem Soc, 2004, 126: 5610-5620.

[46] Nyman M, Alam T M, Bonhomme F, Rodriguez M A, Frazer C S, Welk M E. Solid-state structures and solution behavior of alkali salts of the $Nb_6O_{19}^{8-}$ Lindqvist Ion. Journal of Cluster Science, 2006, 17 (2): 197-219.

[47] Dale B W, Pope M T. The heteropoly-12-niobomanganate (Ⅳ) anion. Chem Commun (London), 1967: 792.

[48] Flynn C M, Jr Stucky G D. Sodium 6-niobo (ethylenediamine) cobaltate (Ⅲ) and its chromate (Ⅲ) analog. Inorg Chem, 1969, 8 (1): 178-180.

[49] Flynn C M, Jr Stucky G D. Heteropolyniobate complexes of manganese (Ⅳ) and nickel (Ⅳ). Inorg Chem, 1969, 8 (2): 332-334.

[50] Flynn C M, Jr Stucky G D. Crystal structure of sodium 12-niobomanganate (Ⅳ), $Na_{12}MnNb_{12}O_{38} \cdot 50H_2O$. Inorg Chem, 1969, 8 (2): 335-344.

[51] Ozeki T, Yamase T, Naruke H, Sasaki Y. Synthesis and structure of dialuminiohexaeuropiopentakis (hexaniobate): a high-nuclearity oxoniobate complex. Inorg Chem, 1994, 33: 409-410.

[52] Naruke H, Yamase T. $Na_8H_{18}[\{Er_3O(OH)_3(H_2O)_3\}_2Al_2(Nb_6O_{19})_5] \cdot 40.5H_2O$. Acta Cryst, 1996, C52: 2655-2660.

[53] Yamase T, Naruke H. Luminescence and energy transfer phenomena in Tb^{3+}/Eu^{3+}-mixed polyoxometallolanthanoates $K_{15}H_3[Tb_{1.4}Eu_{1.6}(H_2O)_3(SbW_9O_{33})(W_5O_{18})_3] \cdot 22.5H_2O$ and $Na_7H_{19}[Tb_{4.3}Eu_{1.7}O_2(OH)_6(H_2O)_6Al_2(Nb_6O_{19})_5] \cdot 47H_2O$. J Phys Chem B, 1999, 102: 8850-8857.

[54] Silva M F P, Cavaleiro A M V, Jesus J D P D. Synthesis and characterisation of a new niobate complex with chromium (Ⅲ). J Coord Chem, 2000, 50: 145-151.

[55] Hegetschweiler K, Finn R C, Rarig R S, Sander J J, Steinhauser S, Wörle M, Zubieta J. Surface complexation of $[Nb_6O_{19}]^{8-}$ with $Ni^{Ⅱ}$: solvothermal synthesis and X-ray structural characterization of two novel heterometallic Ni-Nb-polyoxometalates. Inorganica

[56] Bontchev R P, Venturini E L, Nyman M. Copper-linked hexaniobate Lindqvist clusters-variations on a theme. Inorg Chem, 2007, 46 (11): 4483-4491.

[57] Wang Jingping, Niu Hongyu, Niu Jingyang. A novel Lindqvist type polyoxoniobate coordinated to four copper complex moieties: $\{Nb_6O_{19}[Cu(2,2'-bipy)]_2[Cu(2,2'-bipy)_2]_2\} \cdot 19H_2O$. Inorg Chem Commun, 2008, 11: 63-65.

[58] Besserguenev A V, Dickman M H, Pope M T. Robust, alkali-stable, triscarbonyl metal derivatives of hexametalate anions, $[M_6O_{19}\{M'(CO)_3\}_n]^{(8-n)-}$ (M=Nb, Ta; M'= Mn, Re; $n=1,2$). Inorg Chem, 2001, 40: 2582-2586.

[59] Lindqvist I, Aronsson B. The structure of the hexatantalate ion in $4K_2O \cdot 3Ta_2O_5 \cdot 16H_2O$. Arkiv Kemi, 1954, 7: 49-53.

[60] Pickhard F, Hartl H. The crystal structures of $K_8Ta_6O_{19} \cdot 16 H_2O$ and $K_7NaTa_6O_{19} \cdot 14H_2O$. Z Anorg Allg Chem, 1997, 623: 1311-1316.

[61] Hartl H, Pickhard F, Emmerling F, Rohr C. Rubidium and caesium compounds with the isopolyanion $[Ta_6O_{19}]^{8-}$——synthesis, crystal structure, thermogravimetric and vibrational spectrocopic analysis of the oxotantalates $A_8[Ta_6O_{19}] \cdot nH_2O$(A=Rb, Cs; $n=0,4,14$). Z Anorg Allg Chem, 2001, 627: 2630-2638.

[62] Abramov P A, Abramova A M, Peresypkina E V, Gushchin A L, Adonin S A, Sokolov M N. New polyoxotantalate salt $Na_8[Ta_6O_{19}] \cdot 24.5H_2O$ and its properties. J Struct Chem, 2011, 52 (5): 1012-1017.

[63] Guo G L, Xu Y Q, Chen B K, Lin Z G, Hu C W. Two novel polyoxotantalates formed by Lindqvist-type hexatantalate and copper-amine complexes. Inorg Chem Commun, 2011, 14: 1448-1451.

[64] Fuchs J, Jahr K F. Über neue polywolframate und-molybdate. Z Naturforschg, 1968, 23b: 1380.

[65] Allcock H R, Bissell E C, Shawl E T. Crystal and molecular structure of a new hexamolybdate-cyclophosphazene complex. Inorg Chem, 1973, 12 (12): 2963-2968.

[66] Che M, Fournier M, Launay J P. The analog of surface molybdenyl ion in Mo/SiO_2 supported catalysts: the isopolyanion $Mo_6O_{19}^{3-}$ studied by EPR and UV-visible spectroscopy. Comparison with other molybdenyl compounds. J Chem Phys, 1979, 71 (4): 1954.

[67] Ginsberg A P. Inorgnic synchesis: Vol 27. New York: John Wiley & Sons, 1990.

[68] Kepert D L. Structures of polyanions. Inorg Chem, 1969, 8 (7): 1556-1558.

[69] Dahlstrom P, Zubieta J. Crystal and molecular structure of the tetrabutylammonium salt of the nonadecaoxo-hexamolybdate (Ⅵ) dianion, $[(n-C_4H_9)_4N]_2[Mo_6O_{19}]$. Cryst Struct Comm, 1982, 11: 463.

[70] Arzoumanian H, Baldy A, Lai R, Odreman A, Metzger J, Pierrot M. An unusual route to the isopolymolybdates: octamolybdate $\beta-[Mo_8O_{26}]^{4-}$ and hexamolybdate $[Mo_6O_{19}]^{2-}$. Reaction of dioxomolybdenum complexes with triphenylphosphonium

ylides. Crystal structures of the salts $[PPh_3CH_2COOEt]_2^+[NH_2Et]_2^+[Mo_8O_{26}]^{4-}$, $[PPh_3CH_2COOEt]_2^+[Mo_6O_{19}]^{2-}$, and $[PPh_3CH_2Ph]_2^+[Mo_6O_{19}]^{2-}$. J Organomet Chem, 1985, 295: 343-352.

[71] Ray C, Bernard F H, Penny P. Carbonyl halides of the group 6 transition metals. XXIX. The crystal and molecular structure of $[Mo(CO)_2(dpmSe)_2Cl]_2[Mo_6O_{19}] \cdot 4CH_3NO_2$ [dpmSe= $Ph_2PCH_2P(Se)Ph_2$]. Aust J Chem, 1988, 41: 1295-1303.

[72] Triki S, Ouahab L, Halet Jean-Francois, Pena O, Padiou J, Grandjean D, Garrigou-Lagrange C, Delhaes P. Preparation and properties of tetrathia-and tetramethyl-tetraselena-fulvalene salts of $[Mo_6O_{19}]^{2-}$ (M = Mo or W)$^+$. J Chem Soc Dalton Trans, 1992: 1217.

[73] Triki S, Ouahab L, Grandjean D. Structures of two bis (ethylenedithio) tetrathiafulvalene hexamolybdate and hexatungstate salts: $(BEDT-TTF)_2Mo_6O_{19}$, M = Mo, W. Acta Cryst, 1991, C47: 645-648.

[74] Rheingold A L, White C B, Haggerty B S. Bis (tetrabutylammonium) nonadecaoxohexamolybdenum (Ⅵ): a second polymorph. Acta Cryst, 1993, C49: 756-758.

[75] Triki S, Ouahab L. 3,3'-Dimethyl-4,4'-diphenyl-2,2',5,5'-tetrathiafulvalenium hexamolybdate, $2(C_{20}H_{16}S_4^+)[Mo_6O_{19}]^{2-}$. Acta Cryst, 1994, C50: 219-221.

[76] Attanasio D, Bellitto C, Bonamico M, Fares V, Imperatori P. Synthesis, crystal structure and spectroscopic studies of new charge-transfer compounds derived from the organic donor molecule tetrathiafulvalene, TTF, and inorganic polyoxoanions $[Mo_6O_{19}]^{2-}$, where M=Mo, W. Gazzetta Chimica Italiana, 1991, 121: 155.

[77] Xu X, You X, Wang X. Synthesis, crystal structure and photochemical study of bis (2,4,6-triphenylpyryllium) hexamolybdate (TPPM) and $[n-Bu_4N]_2[Mo_6O_{19}] \cdot$ (TBAM). Acta Chemica Scandinavica, 1996, 50: 1-5.

[78] Hoppe S, Stark J L, Whitmire K H. Bis [bis(triphenylphosphine) iminium] hexamolybdate. Acta Cryst, 1997, C53: 68-70.

[79] Xu X, You X. Potential molecular materials based on organic-inorganic charge-transfer salts derived from $[Mo_6O_{19}]^{2-}$ isopolyoxoanion and hemicyanine dyes: synthesis, spectra properties and X-ray structure. Polyhedron, 1995, 14 (13-14): 1815-1824.

[80] Wang W, Xu L, Gao G, Wei Y. Tris (1,10-phenanthroline) nickel (Ⅱ) hexamolybdate. Acta Cryst, 2005, E61: m813-m815.

[81] Wu D, Wang S, Lin X, Lu C, Zhuang H. Bis (tetramethylammonium) nonadecaoxohexamolybdenum (Ⅵ) monohydrate. Acta Cryst, 2000, C56: e55-e56.

[82] Strukan N, Cindric M, Devcic M, Giester G, Mamenar B. Bis (tetramethylammonium) hexamolybdate hydrate, $[(CH_3)_4N]_2[Mo_6O_{19}] \cdot H_2O$. Acta Cryst, 2000, C56: e278-e279.

[83] Wang X, Guo Y, Wang E, Duan L, Xu X, Hu C. Synthesis and crystal structure of a new 3D supramolecular network based on rare earth dimer and polyoxometalate: $[Y_2(DNBA)_2(OAc)_2(DMF)_6][Mo_6O_{19}]$(DNBA=3,5-dinitrobenzoate). J Mol Struct,

2004, 691: 171-180.

[84] Sa R, Lu X, Liu B, Wang J. Synthesis, crystal structure, and spectrum characterization of [Na(DB$_{18}$C$_6$)(CH$_3$CN)]$_2$Mo$_6$O$_{19}$. Chinese J Inorg Chem, 2005, 21 (12): 1843-1846.

[85] Veya P L, Kochi J K. Structural and spectral characterization of novel charge-transfer salts of polyoxometalates and the cationic ferrocenyl donor. J Organomet Chem, 1995, 488: C4-C8.

[86] Bernstein S N, Dunbar K R. Novel strategies for the synthesis and crystallization of electrophilic dinuclear cations: solution and solid-state properties of [Re$_2$(NCCH$_3$)$_{10}$][Mo$_6$O$_{19}$]$_2$. Angew Chem Int Ed Engl, 1992, 31 (10): 1360-1362.

[87] Li Y, Hao N, Wang E, Yuan M, Hu C, Hu Ni, Jia H. New high-dimensional networks based on polyoxometalate and crown ether building blocks. Inorg Chem, 2003, 42 (8): 2729-2735.

[88] Wang X, Guo Y, Li Y, Wang E, Hu C, Hu N. Novel polyoxometalate-templated, 3-D supramolecular networks based on lanthanide dimers: synthesis, structure, and fluorescent properties of [Ln$_2$(DNBA)$_4$(DMF)$_8$][Mo$_6$O$_{19}$](DNBA=3,5-dinitrobenzoate). Inorg Chem, 2003, 42: 4135-4140.

[89] Rocchiccioli-Deltcheff C, Thouvenot R, Fouassier M. Vibrational investigations of polyoxometalates. 1. Valence force field of [Mo$_6$O$_{19}$]$^{2-}$ based on total isotopic substitution (^{18}O, ^{92}Mo, ^{100}Mo). Inorg Chem, 1982, 21: 30-35.

[90] Bridgeman A J, Cavigliasso G. Density functional study of the vibrational frequencies of Lindqvist polyanions. Chem Phys, 2002, 279: 143-159.

[91] Peng Z. Rational synthesis of covalently bonded organic-inorganic hybrids. Angew Chem Int Ed, 2004, 43: 930-935.

[92] Gouzerh P, Proust A. Main-group element, organic, and organometallic derivatives of polyoxometalates. Chem Rev, 1998, 98 (1): 77-112.

[93] Hsieh T, Zubieta J A. Synthesis and characterization of oxomolybdate clusters containing coordinatively bound organo-diazenido units: the crystal and molecular structure of the hexanuclear diazenido-oxomolybdate, [Bu$_4$N]$_3$[Mo$_6$O$_{18}$(NNC$_6$H$_5$)]. Polyhedron, 1986, 5 (10): 1655-1657.

[94] Bank S, Liu S, Shaikh S N, Sun X, Zubieta J, Ellis P D. ^{95}Mo NMR studies of (aryldiazenido)-and (organohydrazido) molybdates. Crystal and molecular structure of [n-Bu$_4$N]$_3$[Mo$_6$O$_{18}$(NNC$_6$F$_5$)]. Inorg Chem, 1988, 27: 3535-3543.

[95] Bustos C, Hasenknopf B, Thouvenot R, Vaissermann J, Proust A, Gouzerh P. Lindqvist-type (aryldiazenido) polyoxomolybdates-synthesis, and structural and spectroscopic characterization of compounds of the type [n-Bu$_4$N]$_3$[Mo$_6$O$_{18}$(N$_2$Ar)]. Eur J Inorg Chem, 2003, 15, 2757-2766.

[96] Li Hai-Lian, You Xiao-Zeng, Xu Xue-Xiang. Synthesis and structure of the hexanuclear

diazenido-oxomolybdate $[(n\text{-}C_4H_9)_4N]_3[Mo_6O_{18}(N_2C_6H_4\text{-}p\text{-}NO_2)]$ synthesis and structure of the hexanuclear diazenido-oxomolybdate $[(n\text{-}C_4H_9)_4N]_3[Mo_6O_{18}(N_2C_6H_4\text{-}p\text{-}NO_2)]$. Chinese J Struct Chem, 1994, 13 (2): 109-112.

[97] Wang Shuang xi, Wang Xin, Zhai Ying li. Synthesis, structure and properties of $(n\text{-}Bu_4N)_3[Mo_6O_{18}N_2Ar(NO_2)_2]$. Chem Res Chin Univ, 1992, 4: 382-387.

[98] Kang H, Zubieta J. Coordination complexes of polyoxomolybdates with a hexanuclear core: synthesis and structural characterization of $(Bu_4^nN)_2[Mo_6O_{18}(NNMePh)]$. J Chem Soc Chem Commun, 1988: 1192.

[99] Kwen H, Young V G, Jr Maatta E A. A diazoalkane derivative of a polyoxometalate: preparation and structure of $[Mo_6O_{18}(NNC(C_6H_4OCH_3)CH_3)]^{2-}$. Angew Chem Int Ed, 1999, 38 (8): 1145-1146.

[100] Du Yuhua, Rheingold A L, Maatta E A. A polyoxometalate incorporating an organoimido ligand: preparation and structure of $[Mo_5O_{18}(MoNC_6H_4CH_3)]^{2-}$. J Am Chem Soc, 1992, 114: 345-346.

[101] Qiu Yunfeng, Xu Lin, Gao Guanggang, Wang Wenju, Li Fengyan. A new arylimido derivative of Polyoxometalate $(Bu_4N)_2[Mo_6O_{17}(NAr)_2][Ar=2,6\text{-}(CH_3)_2C_6H_3]$: synthesis, structure and physicochemical properties. Inorgan Chim Acta, 2006, 359: 451-458.

[102] Zhu Yi, Xiao Zicheng, Ge Ning, Wang Na, Wei Yongge, Wang Yuan. Naphthyl amines as novel organoimido ligands for design of POM-based organic-inorganic hybrids: synthesis, structural characterization and supramolectlar assembly of $(Bu_4N)_2[Mo_6O_{18}N(Naph\text{-}1)]$. Cryst Grow Des, 2006, 6 (7): 1620-1625.

[103] Hao Jian, Ruhlmann L, Zhu Yulin, Li Qiang, Wei Yongge. Naphthylimido-substituted hexamolybdate: preparation, crystal structures, solvent effects and optical properties of three polymorphs. Inorg Chem, 2007, 46: 4960-4967.

[104] Li Qiang, Wu Pingfan, Wei Yongge, Xia Yun, Wang Yuan, Guo Hongyou. Organic-inorganic hybrids: preparation and structural characterization of $(Bu_4N)_2[Mo_6O_{17}(NAr)_2]$ and $(Bu_4N)_2[Mo_6O_{18}(NAr)](Ar=o\text{-}CH_3C_6H_4)$. Z Anorg Allg Chem, 2005, 631: 773-779.

[105] Proust A, Thouvenot R, Chaussade M, Robert F, Gouzerh P. Phenylimido derivatives of $[Mo_6O_{19}]^{2-}$: syntheses, X-ray structures, vibrational, electrochemical, [95]Mo and ^{14}N NMR studies. Inorg Chim Acta, 1994, 224: 81-95.

[106] Strong J B, Ostrander R, Rheingold A L, Maatta E A. Ensheathing a polyoxometalate: convenient systematic introduction of organoimido ligands at terminal oxo sites in $[Mo_6O_{19}]^{2-}$. J Am Chem Soc, 1994, 116: 3601-3602.

[107] Clegg W, Errington R J, Fraser K A, Holmes S A, Schäfer A. Functionalisation of $[Mo_6O_{19}]^{2-}$ with aromatic amines: synthesis and structure of a hexamolybdate building block with linear difunctionality. J Chem Soc Chem Commun, 1995: 455.

[108] Strong J B, Haggerty B S, Rheingold A L, Maatta E A. A superoctahedral complex

derived from a polyoxometalate: the hexakis (arylimido) hexamolybdate anion [Mo_6 $(NAr)_6O_{13}H$]. Chem Commun, 1997: 1137.

[109] Moore A R, Kwen H, Beatty A M, Maatta E A. Organoimido-polyoxometalates as polymer pendants. Chem Commun, 2000: 1793-1794.

[110] Strong J B, Yap G P A, Ostrander R, Liable-Sands L M, Rheingold A L, Thouvenot R, Gouzerh P, Maatta E A. A new class of functionalized polyoxometalates: synthetic, structural, spectroscopic, and electrochemical studies of organoimido derivatives of [Mo_6O_{19}]$^{2-}$. J Am Chem Soc, 2000, 122: 639-649.

[111] Wei Yongge, Xu Bubin, Barnes C L, Peng Zhonghua. An efficient and convenient reaction protocol to organoimido derivatives of polyoxometalates. J Am Chem Soc, 2001, 123: 4083-4084.

[112] Roesner R A, McGrath S C, Brockman J T, Moll J D, West D X, Swearingen J K, Castineiras A. Mono-and di-functional aromatic amines with p-alkoxy substituents as novel arylimido ligands for the hexamolybdate ion. Inorg Chim Acta, 2003, 342: 37-47.

[113] Xu Bubin, Peng Zhonghua, Wei Yongge, Powell D R. Polyoxometalates covalently bonded with terpyridine ligands. Chem Commun, 2003: 2562-2563.

[114] Li Qiang, Wu Pingfan, Wei Yongge, Wang Yuan, Wang Ping, Guo Hongyou. Synthesis, structure and supramolecular assembly in the crystalline state of a bifuncionalized arylimido derivative of hexamolybdate. Inorg Chem Commun, 2004, 7: 524-527.

[115] Kwen H, Beatty A M, Maatta E A. A p-cyanophenylimido hexamolybdate: preparation and structure of [(n-C_4H_9)$_4$N]$_2$, [Mo_6O_{18} (N-p-C_6H_4CN)]. C R Chimie, 2005, 8: 1025-1028.

[116] Xu Bubin, Lu Meng, Kang J, Wang Degang, Brown J, Peng Zhonghua. Synthesis and optical properties of conjugated polymers containing polyoxometalate clusters as side-chain pendants. Chem Mater, 2005, 17: 2841-2851.

[117] Kang J, Xu Bubin, Peng Zhonghua, Zhu Xiaodong, Wei Yongge, Powell D R. Molecular and polymeric hybrids based on covalently linked polyoxometalates and transition-metal complexes. Angew Chem Int Ed, 2005, 44: 6902-6905.

[118] Lu Meng, Xie Baohan, Kang J, Chen Fang-Chung, Yang Yang, Peng Zhonghua. Synthesis of main-chain polyoxometalate-containing hybrid polymers and their applications in photovoltaic cells. Chem Mater, 2005, 17: 402-408.

[119] Lu Meng, Kang J, Wang Degang, Peng Zhonghua. Enantiopure 1,1′-binaphthyl-based polyoxometalate-containing molecular hybrids. Inorg Chem, 2005, 44: 7711-7713.

[120] Bar-Nahum I, Narasimhulu K V, Weiner L, Neumann R. Phenanthroline-polyoxometalate hybrid compounds and the observation of intramolecular charge transfer. Inorg Chem, 2005, 44: 4900-4902.

[121] Xia Yun, Wei Yongge, Guo Hongyou. Monosubstituted organoimido derivative of the polyoxometalate: synthesis of (n-Bu_4N)$_2$[$Mo_6O_{18}(NAr)$] (Ar=o-$CH_3OC_6H_4$). Acta

Chim Sin, 2005, 63 (20): 1931-1935.

[122] Qin Chao, Wang Xinlong, Xu Lin, Wei Yongge. A linear bifunctionalized organoimido derivative of hexamolybdate: convenient synthesis and crystal structure. Inorg Chem Commun, 2005, 8: 751-754.

[123] Xia Yun, Wu Pingfan, Wei Yongge, Wang Yuan, Guo Hongyou. Synthesis, crystal structure, and optical properties of a polyoxometalate-based inorganic-organic hybrid solid, $(n\text{-Bu}_4\text{N})_2[\text{Mo}_6\text{O}_{17}(\text{NAr})_2]$ $(\text{Ar}=o\text{-CH}_3\text{OC}_6\text{H}_4)$. Cryst Growth Des, 2006, 6 (1): 253-257.

[124] Li Qiang, Wei Yongge, Hao Jian, Zhu Yulin, Wang Longsheng. Unexpected C=C bond formation via doubly dehydrogenative coupling of two saturated sp^3 C—H bonds activated with a polymolybdate. J Am Chem Soc, 2007, 129: 5810-5811.

[125] Gouzerh P, Jeannin Y, Proust A, Robert F. Two novel polyoxomolybdates containing the $(\text{MoNO})^{3+}$ unit: $[\text{Mo}_5\text{Na}(\text{NO})\text{O}_{13}(\text{OCH}_3)_4]^{2-}$ and $[\text{Mo}_6\text{O}_{18}(\text{NO})]^{3-}$. Angew Chem Int Ed Engl, 1989, 28 (10): 1363.

[126] Proust A, Thouvenot R, Robert F, Gouzerh P. Molybdenum oxo nitrosyl complexes. 2. ^{95}Mo NMR studies of defect and complete Lindqvist-type derivatives. Crystal and molecular structure of $(n\text{-Bu}_4\text{N})_2[\text{Mo}_6\text{O}_{17}(\text{OCH}_3)(\text{NO})]$. Inorg Chem, 1993, 32: 5299-5304.

[127] Bottomley F, Chen Jinhua. Organometalic oxides: oxidation of $[(\eta\text{-C}_5\text{Me}_5)\text{Mo}(\text{CO})_2]_2$ with O_2 to form syn-$[(\eta\text{-C}_5\text{Me}_5)\text{MoCl}]_2(\mu\text{-Cl})_2(\mu\text{-O})$, syn-$[(\eta\text{-C}_5\text{Me}_5)\text{MoCl}]_2(\mu\text{-Cl})(\mu\text{-CO}_3\text{H})(\mu\text{-O})$, and $[\text{C}_5\text{Me}_5\text{O}][(\eta\text{-C}_5\text{Me}_5)\text{Mo}_6\text{O}_{18}]$. Organometallics, 1992, 11: 3404-3411.

[128] Proust A, Thouvenot R, Herson P. revisiting the synthesis of $[\text{Mo}_6(\eta^5\text{-C}_5\text{Me}_5)\text{O}_{18}]^-$ X-ray structural analysis, UV-visible, electrochemical and multinuclear NMR characterization. J Chem Soc Dalton Trans, 1999: 51-55.

[129] Fuchs J. Polyanions of novel structural types. Z Naturforch, 1973, 28b: 389-404.

[130] Chaudron M G. Chimie physique-nouvelle préparation et properieties de l'ion hexatungstique $\text{W}_6\text{O}_{19}^{2-}$. Note (*) de MM. Michel Boyer er Bernard Le Meur, présentée. C R Acad Sc Paris, 1975, 281: 59-62.

[131] Himeno S, Kitazumi I. Capillary electrophoretic study on the formation and transformation of isopolyoxotungstates in aqueous and aqueous-CH_3CN media. Inorg Chim Acta, 2003, 355: 81-86.

[132] Fuchs J, Freiwald W, Hartl H. Neubestimmung der kristallstruktur von tetrabutylammoniumhexawolframat. Acta Cryst, 1978, B34: 1764-1770.

[133] Larue W A, Liu A T, Filippo J S. Preparation and structure of heptakis (tert-butylisocyanide) tungsten (II) hexatungstate, $[(t\text{-C}_4\text{H}_9\text{NC})_7\text{W}][\text{W}_6\text{O}_{19}]$. Inorg Chem, 1980, 19: 315-320.

[134] Triki S, Ouahab L, Grandjean D. Structure of bis (2,3,6,7-tetramethyl-1,4,5,8-te-

trathiafulvalenium) hexatungstate and hexamolybdate. (TMTTF)$_2$M$_6$O$_{19}$, M = W. Acta Cryst, 1991, C47: 1371-1373.

[135] Hou H W, Ye X R, Xin X Q. Tetrabutylammonium di-μ-iodo-bis (iodoargentate) hexatungstate, [^{n}Bu$_4$N]$_4$[Ag$_2$I$_4$][W$_6$O$_{19}$]. Acta Cryst, 1995, C51: 2013-2015.

[136] Parvez M, Boorman P M, Langdon N. Tetraphenylphosphonium hexatungstate (Ⅵ) acetonitrile solvate. Acta Cryst, 1998, C54: 608-609.

[137] Baker P K, Drew M G B, Meehan M M. Cationic linear triphos, L [PhP(CH$_2$CH$_2$PPh$_2$)$_2$], alkyne complexes of molybdenum (Ⅱ) and tungsten (Ⅱ); crystal structures of [WI(CO)(L-P, P′, P″) (η^2-MeC$_2$Me)]$_2$[W$_6$O$_{19}$], [WI(CO)(L-P, P′, P″)(η^2-MeC$_2$Me)][BPh$_4$] and [MoBr$_2$(O){Ph$_2$P(CH$_2$)$_2$PPh(CH$_2$)$_2$POPh$_2$-P,P′,P″,O}]. J Chem Soc Dalton Trans, 1999: 765-771.

[138] Lin Xin rong, Chen Ming qin, Liu Hui zhang, Xu Jin yu, Li Pei, Jin Song lin, Xie Gao yang. Synthesis and crystal structure of di-p-methylpyridinium-p-methylpyridinidine hexatungstate. Chem Res Chi Univ, 1999, 15 (3): 205-210.

[139] Meng F X, Liu K, Chen Y G. Hydrothermal synthesis, crystal structure and spectral characterization of a new copper isopolytungstate: [Cu(phen)$_3$][W$_6$O$_{19}$]. Chi J Struct Chem, 2006, 25 (7): 837-843.

[140] Wang X L, Bi Y F, Liu G C, Lin H Y. Synthesis and structure of a hexatungstate-templated 3D supramolecular network [La$_2$(DNBA)$_4$(DMF)$_8$][W$_6$O$_{19}$]. Chi J Inorg Chem, 2006, 22 (5): 957-962.

[141] Bhattacharyya R, Biswas S, Armstrong J, Holt E M. New and general route to the synthesis of oxopolymetalatesvia peroxometalates in aqueous medium: synthesis and crystal and molecular structure of (PPN)$_2$[W$_6$O$_{19}$] (PPN=bis (triphenylphosphine) nitrogen (1+) cation). Inorg Chem, 989, 28: 4297-4300.

[142] Reinoso S, Dickman M H, Kortz U. A novel hexatungstate fragment stabilized by dimethyltin groups: [{(CH$_3$)$_2$Sn}$_2$(W$_6$O$_{22}$)]$^{4-}$. Inorg Chem, 2006, 45: 10422-10424.

[143] Wang Jing ping, Du Xiao di, Niu Jing yang. A novel 1D organic-inorganic hybrid based on alternating heteropolyanions [GeMo$_{12}$O$_{40}$]$^{4-}$ and isopolyanions [Mo$_6$O$_{22}$]$^{8-}$. J Solid State Chem, 2006, 179: 3260-3264.

[144] Mohs T R, Yap G P A, Rheingold A L, Maatta E A. An organoimido derivative of the hexatungstate cluster: preparation and structure of [W$_6$O$_{18}$(NAr)]$^{2-}$ (Ar=2,6-(i-Pr)$_2$C$_6$H$_3$). Inorg Chem, 1995, 34: 9-10.

[145] Yamase T, Pope M T. Polyoxometalate chemistry for nano-composite design. Amsterdam: Kluwer Academic/Plenum Publishers, 2002.

[146] Proust A, Thouvenot R, Roh S G, Yoo J K, Gouzerh P. Lindqvist-type oxo nitrosyl complexes. Syntheses, vibrational, multinuclear magnetic resonance (^{14}N, ^{17}O, ^{95}Mo, and ^{183}W), and electrochemical studies of [M$_5$O$_{18}${M′(NO)}]$^{3-}$ anions (M, M′=Mo, W). Inorg Chem, 1995, 34: 4106-4112.

[147] Che T M, Day V W, Francesconi L C, Fredrich M F, Klemperer W G. Synthesis and structure of the $[(\eta^5\text{-}C_5H_5)Ti(Mo_5O_{18})]^{3-}$ and $[(\eta^5\text{-}C_5H_5)Ti(W_5O_{18})]^{3-}$ anions. Inorg Chem, 1985, 24: 4055-4062.

[148] Harper J R, Rheingold A L. Arsaoxanes as reversible, ligating oxygen-transfer agents in the synthesis of neutral metal-oxo clusters. The X-ray structures of $Cp_2^*W_6O_{17}$ and $Cp_6^*Mo_8O_{16}$. J Am Chem Soc, 1990, 112: 4037-4038.

第7章

七金属氧簇

能够形成七金属氧簇的元素比较少,目前分离出来的只有七钼氧簇和七钨氧簇。尽管七铌氧簇可以作为构筑块存在于化合物中,但具有独立结构的七铌氧簇尚未分离出来。

7.1 七铌氧簇

有关铌多金属氧簇的报道主要集中在 Lindqvist 结构的六铌氧簇[1-4]。尽管目前尚未有独立结构的七铌氧簇报道,但由七铌氧簇作为构筑块形成的化合物已有多个[5,6]。

七铌氧簇结构如图 7-1 所示,它是在具有 Lindqvist 结构的六铌氧簇的一个面上,通过三个桥氧与第七个铌原子相连,第七个铌原子上有三个端基氧原子,形成组成为 $[Nb_7O_{22}]^{9-}$ 的构筑块,由于该构筑块具有较高的

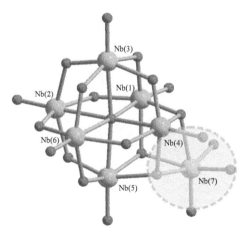

图 7-1 多铌氧簇阴离子中存在的七铌氧簇构筑块

电荷密度，而且其中的一个铌原子上有三个端基氧原子，因此它的配位能力很强。

7.2 七钼氧簇

MoO_4^{2-} 在水溶液中浓度大于 10^{-3} mol/L 时，在一定的酸度下，可直接聚合为七钼氧簇阴离子[7-12]。由三氧化钼与碱或有机胺类在水溶液中回流，也可以生成含七钼氧簇阴离子的化合物[13-18]。

7.2.1 合成

工业上用氨水浸取三氧化钼制七钼酸铵：

$$7MoO_3 + 6NH_3 \cdot H_2O \longrightarrow (NH_4)_6Mo_7O_{24} + 3H_2O$$

七钼酸钾的制备可按如下步骤进行：按物质的量比称取三氧化钼和氢氧化钾，溶于适量水中，调 pH 值到 6，缓慢蒸发，可以得到分子组成为 $K_6[Mo_7O_{24}] \cdot 4H_2O$ 的晶体[19]。

$$2KOH + MoO_3 \longrightarrow K_2MoO_4 + H_2O$$
$$K_2MoO_4 + H^+ \longrightarrow K_6Mo_7O_{24} + H_2O + K^+$$

7.2.2 谱学表征

图 7-2 为 $(NH_4)_6[Mo_7O_{24}] \cdot 4H_2O$ 红外光谱，由谱图可知，阴离子的特征吸收峰出现在 $1000 \sim 350 cm^{-1}$ 范围内。其端基氧的振动峰出现在 $900 cm^{-1}$ 和 $940 cm^{-1}$[20]。

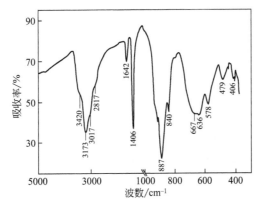

图 7-2　$(NH_4)_6[Mo_7O_{24}] \cdot 4H_2O$ 红外光谱

7.2.3 晶体结构

1950 年 Lindqvist 报道了七钼酸铵 $(NH_4)_6[Mo_7O_{24}] \cdot 4H_2O$ 的晶体结构[19]，这是第一次报道的有关七钼氧簇的结构，但他仅确定了阴离子中七个钼原子的位置，而没有确定氧原子的位置，但他基于假定合理的钼氧键长和共边八面体构型，提出了整个阴离子最可能的构型。后来又有多位学者对七钼氧簇的结构进行了修正并确定了阴离子中氧原子的位置[13,21-23]。阴离子结构如图 7-3 所示，阴离子由七个共边 MoO_6 八面体构成，具有 2mm 对称性 C_{2v} 点群。

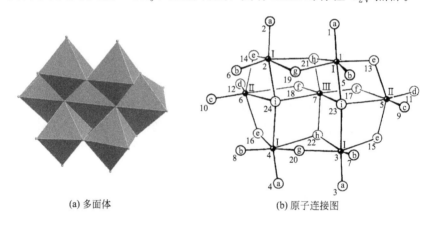

(a) 多面体 (b) 原子连接图

图 7-3 七钼氧簇阴离子结构

根据阴离子中氧原子的键合情况，七钼氧簇阴离子中的氧原子可分为四类：①仅与一个钼原子相连的端基氧原子（O_t），阴离子中有 12 个端基氧原子；②与两个钼原子相连的二重桥氧（O_b），阴离子中共有 8 个二重桥氧原子；③与三个钼原子相连的三重桥氧原子（O_c），阴离子中共有 2 个三重桥氧原子；④与四个钼原子相连的四重桥氧原子（O_d），阴离子中共有 2 个四重桥氧原子。根据钼原子所处的位置，阴离子中的钼氧八面体 MoO_6 可分为三层，中间一层 3 个钼氧八面体几乎位于一条直线上，其上面和下面各与两个共边的钼氧八面体相连。根据键合氧原子类型的不同，阴离子中的钼原子可分为三种类型：①Mo^a，与四种类型的氧原子键合的钼原子，位于阴离子中上层的两个钼原子和下层的两个钼原子属于该类型；②Mo^b，键合端氧、二重桥氧和四重桥氧原子的钼原子，位于中层两端的钼原子属于该类型；③Mo^c，键合二重桥氧、三重桥氧和四重桥氧原子的钼原子，位于阴离子中心的钼原子属于该类型。由于位于阴离子中心的钼原子上没有端基氧，因此，由中心钼原子与桥氧原子形成键的键长与该氧原子同其他钼原子成键的键长差别较大。表 7-1 列出了部分化合物中七钼氧簇阴离子的键长。

表 7-1　部分化合物中七钼氧簇阴离子键长　　　　　　　单位：Å

化合物	Mo—O_t	Mo^a—O_b	Mo^a—O_c	Mo^a—O_d	Mo^b—O_b	Mo^b—O_d	Mo^c—O_b	Mo^c—O_c	Mo^c—O_d	文献
$(NH_4)_6[Mo_7O_{24}] \cdot 4H_2O$	1.724	1.956	2.179		2.165	2.155	1.754	1.903	2.264	[23]
$[C_6H_{18}N_2]_6Mo_7O_{24} \cdot 4H_2O$	1.721	1.963	2.282	2.168	2.123	2.171	1.737	1.898	2.275	[18]
$(NH_4)_6[Mo_7O_{24}] \cdot 4H_2O$	1.728	1.955	2.29	2.18	2.123	2.16	1.744	1.90	2.247	[13]
$K_6[Mo_7O_{24}] \cdot 4H_2O$	1.74	1.93	2.22	2.18	2.13	2.16	1.68	1.92	2.26	[13]
$(C_4H_{16}N_3)_6[Mo_7O_{24}] \cdot 4H_2O$	1.723	1.964	2.27	2.16	2.14	2.16	1.74	1.90	2.28	[14]
$(C_4H_{16}N_3)_2[Mo_7O_{24}]$	1.72	1.95	2.26	2.17	2.11	2.17	1.74	1.91	2.25	[24]
$(C_5H_{15}N)[Mo_7O_{24}]$	1.72	1.96	2.24	2.17	2.12	2.16	1.74	1.91	2.28	[25]
$(C_3H_{10}N)[Mo_7O_{24}]$	1.71	1.96	2.25	2.17	2.12	2.17	1.73	1.90	2.29	[15]

7.2.4　性质研究

由于七钼氧簇阴离子形成的化合物在多个领域具有重要应用，人们对它所形成化合物的性质进行了广泛研究。

（1）热性质

有关七钼酸铵 $(NH_4)_6[Mo_7O_{24}] \cdot 4H_2O$ 热性质的研究有很多报道[26-48]。人们应用热重、差热、红外光谱、拉曼光谱和 X 射线衍射等多种手段研究了 $(NH_4)_6[Mo_7O_{24}] \cdot 4H_2O$ 在不同的受热条件下的分解过程及分解产物[28-32,34,35,49]。T. Ressler 等采用原位质谱和 X 射线跟踪的方法，研究了在含有 20%氧气的氮气中 $(NH_4)_6[Mo_7O_{24}] \cdot 4H_2O$ 的受热过程[49]，图 7-4 为加热速度 5K/min、气体流速 100mL/min 条件下的热谱图。化合物的热失重分为四步，由此可以判断化合物的热分解分为四步，其分解温度分别为 320K、460K、

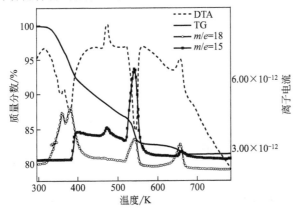

图 7-4　$(NH_4)_6[Mo_7O_{24}] \cdot 4H_2O$ 热谱图

530K 和 635K。第三步失重,530K 时失重率达到 17.22%;第四步失重,635K 时总失重率达到 18.61%,与 $(NH_4)_6[Mo_7O_{24}]\cdot 4H_2O$ 最终分解产物为 α-MoO_3 时理论失重率 18.48% 一致。第一步失重有两个可分辨的吸热峰,第三步失重也表现为吸热,而第二步和第四步失重表现为放热。从原位质谱可以看出,除第一步失重有失水外,其他各步失重均有氨和水的失去。

升温速度也会对七钼酸铵的失重过程产生影响。图 7-5 给出了在不同的升温速度下化合物的热重曲线。将升温速度由 5K/min 降为 0.5K/min,TG 曲线的形状不改变但失重温度降低,当升温速度由 5K/min 提高为 15K/min 时,TG 曲线的形状发生了改变,第二步失重变得明显,除最后一步外,各步的失重温度都升高。改变气流速度,也能对化合物的热重曲线产生影响。

图 7-5 升温速度对 $(NH_4)_6[Mo_7O_{24}]\cdot 4H_2O$ 失重的影响

(2) 光敏性

七钼氧簇阴离子可以作为电子受体与电子给体结合,形成电荷转移化合物,在适当的光源激发下,可以发生给体到受体的电荷转移,进而显示出光敏性[18,50-53]。

研究表明,七钼氧簇阴离子形成的电荷转移化合物的变色过程基于如下方式进行:

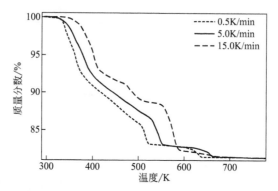

光照变色后化合物的 ESR 谱(图 7-6)上,出现了 Mo^{5+} 的顺磁信号,表明在光照的过程中阴离子得到了电子被还原。

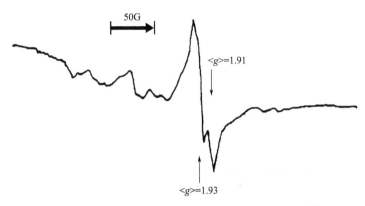

图 7-6 七钼氧簇电荷转移化合物光照变色后 ESR 谱

七钼氧簇阴离子在有机溶液中在光激发下也可与有机分子发生电荷转移，其中的 Mo^{VI} 被还原为 Mo^{V}，而表现出光敏性[54-56]。

7.2.5 相关化合物

由六钼酸铵与 cis-$Ru(DMSO)_4Cl_2$ 和 cis-$Os(DMSO)_4Cl_2$ 反应，可以得到分子组成为 $(NH_4)_4[M(II)(DMSO)_3Mo_7O_{24}] \cdot 6.5H_2O$ 的化合物[57]，该化合物阴离子结构如图 7-7 所示。七个钼原子的排列方式与七钼酸铵中阴离子的排布方式差别很大，而与八钼氧簇中的 β-$Mo_8O_{26}^{4-}$（图 8-9）钼原子的排列方式很像，其中的一个钼原子被 $Ru(II)$[或 $Os(II)$]配离子取代，形成一种新颖结构的七钼氧簇。

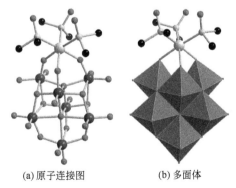

(a) 原子连接图　　(b) 多面体

图 7-7 化合物 $(NH_4)_4[M(II)(DMSO)_3Mo_7O_{24}] \cdot 6.5H_2O$ 结构

由六钼酸铵与稀土硝酸盐或氯化物反应，可以得到组成为 $[Ln_4(MoO_4)(H_2O)_{16}(Mo_7O_{24})_4]^{14-}$（$Ln = La^{III}, Ce^{III}, Pr^{III}, Sm^{III}, Gd^{III}$）的化合物[58]。阴离子结构如图 7-8 所示，阴离子中的四个稀土原子为九配位，三加冠三棱柱构型，其中的一个配位氧原子来源于 MoO_4^{2-} 基团，四个氧原子来源于三个 $Mo_7O_{24}^{6-}$

基团，另外四个配位氧原子来源于配位水，每一个七钼氧簇阴离子相当于一个四齿配体，同时与三个稀土离子配位，位于 $[Ln_4(MoO_4)(H_2O)_{16}]^{10+}$ 基团的四周。

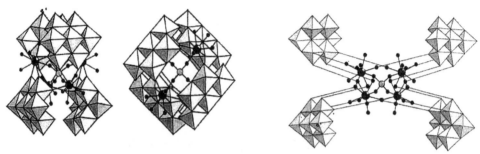

图 7-8 $[Ln_4(MoO_4)(H_2O)_{16}(Mo_7O_{24})_4]^{14-}$ 多面体结构

7.3 七钨氧簇

七钨氧簇阴离子又称为仲钨酸 A 阴离子，它和十二钨氧簇阴离子仲钨酸 B 阴离子在溶液中存在平衡：

$$12[W_7O_{24}]^{6-} + 2H^+ + 6H_2O \rightleftharpoons 7[W_{12}O_{42}H_2]^{10-}$$

7.3.1 合成

早在 1863 年，Marignac 就指出，$3Na_2O \cdot 7WO_3 \cdot 21H_2O$（$\equiv Na_6W_7O_{24} \cdot 21H_2O$）的三斜晶体可以从长时间煮沸的仲钨酸钠的溶液中离析[59]。后来的研究工作表明，从均相酸化到 pH=5.96 的钨酸盐溶液中可以得到 $Na_6W_7O_{24} \cdot 14H_2O$ 的晶体，该晶体与相应的钼氧簇形成的盐同晶型[60]。

图 7-9 化合物 $Cs_2(CN_3H_6)_3$ $[(C_2H_5)_2NH_2][W_7O_{22}(O_2)_2] \cdot 3H_2O$ 中有机分子及阴离子结构

7.3.2 结构

七钨氧簇阴离子具有和七钼氧簇阴离子相同的结构。

7.3.3 相关化合物

Hashimoto 等人报道了分子组成为 $Cs_2(CN_3H_6)_3[(C_2H_5)_2NH_2][W_7O_{22}(O_2)_2] \cdot 3H_2O$ 的化合物[61]。七钨氧簇阴离子结构如图 7-9 所示。

参考文献

[1] Charles M Flynn Jr, Galen D Stucky. Heteropolyniobate complexes of manganese (Ⅳ) and nickel (Ⅳ). Inorg Chem, 1969, 8 (2): 332-334.

[2] Besserguenev A V, Dickman M H, Pope M T. Robust, alkali-stable, triscarbonyl metal derivatives of hexametalate anions, $[M_6O_{19}\{M'(CO)_3\}_n]^{(8-n)-}$ (M=Nb, Ta; M'=Mn, Re; n=1,2). Inorg Chem, 2001, 40 (11): 2582-2586.

[3] Laurencin D, Thorvenot R, Boubekeur K, Proust A. Synthesis and reactivity of $\{Ru(p\text{-}cymene)\}^{2+}$ derivatives of $[Nb_6O_{19}]^{8-}$: a rational approach towards fluxional organometallic derivatives of polyoxometalates. Dalton Trans, 2007: 1334-1345.

[4] Niu Jingyang, Fu Xiao, Zhao Junwei, Li Suzhi, Ma Pengtao, Wang Jingping. Two-dimensional polyoxoniobates constructed from Lindqvist-type hexaniobates functionalized by mixed ligands. Cryst Growth Des, 2010, 10 (7): 3110-3119.

[5] Bontchev R P, Nyman M. Evolution of polyoxoniobate cluster anions. Angew Chem Int Ed, 2005, 45: 6670-6672.

[6] Niu Jingyang, Ma Pengtao, Niu Hongyu, Li Jie, Zhao Junwei, Song You, Wang Jingping. Giant polyniobate clusters based on $[Nb_7O_{22}]^{9-}$ units derived from a Nb_6O_{19} precursor. Chem Eur J, 2007, 13, 43: 4466-4470.

[7] Aveston J, Anacker E W, Johnson J S. Hydrolysis of molybdenum (Ⅵ). Ultracentrifugation, acidity measurements, and Raman spectra of polymolybdates. Inorg Chem, 1964, 3 (5): 735-746.

[8] Honig D S, Kustin K. Relaxation spectra of molybdate polymers in aqueous solution: temperature-jump studies. Inorg Chem, 1972, 11 (1): 65-71.

[9] Brown P L, Shying M E, Syla R N. The hydrolysis of metal ions. Part 10. Kinetic and equilibrium. Measurements of molybdenum (Ⅵ). J Chem Soc Dalton Trans, 1987: 2149-2157.

[10] Cruywagen J J, Heyns J B B. Equilibria and UV spectra of mono-and polynuclear molybdenum (Ⅵ) species. Inorg Chem, 1987, 26: 2569-2572.

[11] Cruywagen J J, Esterhuysen M W, Heyns J B B. The isolation and X-ray characterization of a new sodium heptamolybdate compound, $Na_7[Mo_7O_{24}]$ OH · $21H_2O$. Inorg Chim Acta, 2003, 348: 205-211.

[12] Kortz U, Pope M T. $Cs_6[Mo_7O_{24}]$ · $7H_2O$. Acta Cryst, 1995, C51: 1717-1719.

[13] Evans H T, Gatehouse B M, Leverett P. Crystal structure of the heptamolybdate (Ⅵ) (paramolybdate) ion, $[Mo_7O_{24}]^{6-}$, in the ammonium and potassium tetrahydrate salts. J C S, Dalton, 1975: 505-514.

[14] Roman P, Luque A, Aranzabe A, Gutierrez-Zorrilla J M. Reactions of MoO_3 with diethylenetriamine (dien): syntheses, solid-state characterization and thermal behavior. Molecular and crystal structure of a second polymorph of $(H_3dien)_2[Mo_7O_{24}]$ ·

4H$_2$O. Polyhedron, 1992, 11 (16): 2027-2038.

[15] Ohashi Y, Yanagi K, Sasada Y, Yamase T. Crystal structure and photochemistry of isopolymolybdates. Ⅰ. The crystal structure of hexakis (propylammonium) heptamolybdate (Ⅵ) trihydrate and hexakis (isopropylammonium) heptamolybdate (Ⅵ) trihydrate. Bull Chem Soc Japan, 1982, 55: 1254.

[16] Román P, Gutiérreź-Zorrilla J M, Martínez-Ripoll M, García-Bilanco S. Synthesis, structure and bonding of 2-aminopyridinium heptamolybdate trihydrate. Transition Met Chem, 1986, 11: 143-150.

[17] Attanasio D, Bonamico M, Fares V, Suber L. Organic-inorganic charge-transfer salts based on the β- [Mo$_8$O$_{26}$]$^{4-}$ isopolyanion: synthesis, properties and X-ray structure. J Chem Soc Dalton Trans, 1992: 2523-2528.

[18] Niu J-Y, You X-Z. Synthesis, crystal structure and photochromism of Tri (N,N,N', N'-tetramethylethylendiammonium) heptamolybdate (Ⅵ) tetrahydrate. Polyhedron, 1996, 15 (5-6): 1003-1008.

[19] Lindqvist I. Crystal-structure investigation of the paramolybdate ion. Arkiv Kemi, 1950, 2: 325-341.

[20] Lyhamn L E, Cyvin S J. Symmetry coordinates and force field analysis of a heptamolybdate, Mo$_7$O$_{24}$, model. Z Naturforsch, 1979, 34a: 867-875.

[21] Shimao E. The structure of Mo$_7$O$_{24}^{6-}$ ion in the crystal of ammonium heptamolybdate tetrahydrate. Bull Chem Soc Japan, 1967, 40: 1609.

[22] Gatehouse B M, Leverett P. The crystal structure of potassium heptamolybdate tetrahydrate. Chem Comm, 1968: 901.

[23] Evans H T Jr. Refined molecular structure of the heptamolybdate and hexamolybdotellurate ions. J Am Chem Soc, 1968, 90 (12): 3275-3276.

[24] Román P, Luque A, Gutiérrez-Zorrilla J M, Zúñiga F J. The crystal structure and bonding of bis (diethylenetriammonium) heptamolybdate (Ⅵ) tetrahydrate. Z Kristallogr, 1990, 190: 249-258.

[25] Romm P, Gutierrez-Zorrilla J M, Luque A, Martinez-Ripoll M. Crystal structure and spectroscopic study of polymolybdates. The crystal structure and bonding of hexakis (n-pentylammonium) heptamolybdate (Ⅵ) trihydrate. J Cryst Spectrosc Res, 1988, 18: 117.

[26] Hegedus A J, Sasvari K, Neugebauer J. Thermo-und röntgenanalytischer beitrag zur reduktion des molybdäntrioxyds und zur oxydation bzw. Nitrierung des molybdäns. Z Anorg Allg Chem, 1957, 293: 56.

[27] Onchi M, Ma E. An application of the omegatron mass spectrometer to thermal decomposition studies. J Phys Chem, 1963, 67: 2240-2241.

[28] Ma E. The thermal decomposition of ammonium polymolybdates. Bull Chem Soc Jpn, 1964, 37 (2): 171.

[29] Ma E. The thermal decomposition of ammonium polymolybdates. Bull Chem Soc Jpn, 1964, 37 (5): 648.

[30] Schwing-Weill M-J. Thermogravimetric study of some nitrogen-containing molybdates. I. Ammonium molybdates. Bull Soc Chim Fr, 1967, 10: 3795-3798.

[31] Kiss A B, Gadó P, Asztalos I, Hegedüs A J. New results concerning the thermal decomposition of ammonium heptamolybdate tetrahydrate. Acta Chim Acad Sci Hung, 1970, 66 (3): 235-249.

[32] Louisy A, Dunoyer J M. Termal decomposition of ammonium heptamolibdate. Bull Soc Chim Fr, 1970, 67: 1390-1400.

[33] Bhatnagar I K, Chakrabarty D K, Biswas A B. Thermal decomposition of ammonium vanadate, ammonium molybdate and ammonium tungstate. Indian J Chem, 1972, 1 (10): 1025-1028.

[34] Hanafi Z M, Khilla M A, Askar M H. The thermal decomposition of ammonium heptamolybdate. Thermochimica Acta, 1981, 45: 221.

[35] Isa K, Ishimura H. Thermal decomposition studies of ammonium heptamolybdate (6−) tetrahydrate by means of high-temperature oscillating X-ray diffraction with a rotating anode type large capacity generator. Bull Chem Soc Jpn, 1981, 54: 3628.

[36] Topic M, Moguš-Milankovic A. A multiple thermal analysis of ammonium heptamolybdate tetrahydrate. Chroatica Chem Acta, 1984, 57 (1): 75-83.

[37] Sharma I B, Batra S. Characterization and thermal investigations of ammonium heptamolybdate. J Therm Anal, 1988, 34: 1273-1281.

[38] Wang J Y. The GC study of the thermal decomposition of ammonium paramolybdate tetrahydrate in a hydrogen atmosphere. Thermochimica Acta, 1990, 158: 183.

[39] Shashkin D P, Shiryaev P A. Kutyrev M Y, Krylov O, V. Peculiarities of the effect of vanadium ions on molybdena developmentunder prereaction conditions. Kin Catal, 1993, 34 (2): 302-306.

[40] Halawy S A, Mohamed M A. Characterization of unsupported molybdenum oxide-cobalt oxide catalysts. J Chem Tech Biotechnol, 1993, 58: 237.

[41] Said A A, Halawy S A. Effects of alkali metal ions on the thermal decomposition of ammonium heptamolybdate tetrahydrate. Effects of alkali metal ions on the thermal decomposition of ammonium heptamolybdate tetrahydrate. J Therm Anal, 1994, 41: 1075-1090.

[42] Said A A. Mutual influences between ammonium heptamolybdate and γ-alumina during their thermal treatments. Thermochimica Acta, 1994, 236: 93-104.

[43] Cabello C I, Botto I L, Thomas H J. Reducibility and thermal behaviour of some Anderson phases. Thermochimica Acta, 1994, 232: 183-189.

[44] Bi H, Li H, Pan W P, Lloyd W G, Davis B H. Thermal studies of $(NH_4)_2Cr_2O_7$, $(NH_4)_2WO_4$, and $(NH_4)_6Mo_7O_{24} \cdot 4H_2O$ deposited on ZrO_2. Thermochimica Acta,

[45] Valmalette J C, Houriet R, Hofmann H, Gavarri J R. Formation of N_2O during the thermal decomposition of ammonium salts $(NH_4)_{(a)}M_xO_y$ (M=V, Cr, Mo and W). Eur J Solid State Inorg Chem, 1997, 34: 317.

[46] Yin Zhoulan, Li Xinhai, Chen Qiyuan. Study on the kinetics of the thermal decompositions of ammonium molybdates. Thermochimica Acta, 2000: 352-353, 107.

[47] Thomazeau C, Martin V, Afanasiev P. Effect of support on the thermal decomposition of $(NH_4)_6Mo_7O_{24} \cdot 4H_2O$ in the inert gas atmosphere. Appl Cat A, 2000, 199: 61-72.

[48] Murugan R, Chang H. Thermo-Raman investigations on thermal decomposition of $(NH_4)_6Mo_7O_{24} \cdot 4H_2O$. J Chem Soc Dalton Trans, 2001, 20: 3125.

[49] Wienold J, Jentoft R E, Ressler T. Structural investigation of the thermal decomposition of ammonium heptamolybdate by in situ XAFS and XRD. Eur J Inorg Chem, 2003: 1058-1071.

[50] Yamase T. Photochemical studies of the alkylammonium molybdates. Part 6. Photoreducible octahedron site of $[Mo_7O_{24}]^{6-}$ as determined by electron spin resonance. J Chem Soc Dalton Trans, 1982: 1987.

[51] Yamase T, Sasaki R, Ikawa T. Photochemical studies of the alkylammoniun molybdates. Part 5. Photolysis in weak acid solutions. J Chem Soc Dalton Trans, 1981: 628.

[52] Yamase T. Photochemical studies of alkylammonium molybdates. Part 9. Structure of diamagnetic blue species involved in the photoredox reaction of $[Mo_7O_{24}]^{6-}$. J Chem Soc Dalton Trans, 1991: 3055-3063.

[53] Yamase T, Kurozumi T. Isopolyanions of molybdenum and tungsten as photocatalysts for hydration of acetylene. Inorg Chim Acta, 1984, 83: L25-L27.

[54] Ward M D, Brazdil J F, Grasselli R K. Photocatalytic alcohol dehydrogenation using ammonlum heptamolybdate. J Phys Chem, 1984, 88: 4210-4213.

[55] Kraut B, Ferraudi G. Intermediates in the early events of $Mo_7O_{24}^{6-}$-catalyzed photodehydrogenations: a picosecond-nanosecond flash-photochemical study. Inorg Chem, 1989, 28: 2692-2694.

[56] Andreev V N, Nikitin S E, Klimov V A, Kozyrev S V, Leshchev D V, Shtel'makh K F. Investigation of photochromic cluster systems based on molybdenum oxides by ESR spectroscopy. Physics of the Solid State, 2001, 43 (4): 755-758.

[57] Khenkin A M, Shimon L J W, Neumann R. Preparation and characterization of new ruthenium and osmium containing polyoxometalates, $[M(DMSO)_3Mo_7O_{24}]^{4-}$ (M=Ru(Ⅱ), Os(Ⅱ)), and their use as catalysts for the aerobic oxidation of alcohols. Inorg Chem, 2003, 42: 3331-3339.

[58] Burgemeister K, Drewes D, Limanski E M, Küper I, Krebs B. Formation of large clusters in the reaction of lanthanide cations with heptamolybdate. Eur J Inorg Chem, 2004:

2690-2694.

[59] Marignac C. Recherches chimiques et cristallographiques sur les tungstates, les fluotungstates et les silicotungstates. Ann Chem Phys, 1863, 69: 41.

[60] Sjolom K, Hedman B. Multicomponent polyanions. Ⅷ. The molecular and crystal structure of $Na_6Mo_7O_{24} \cdot 14H_2O$, a compound containing sodium-coordinated heptamolybdate anions. Acta Chem Scand, 1973, 27: 3673.

[61] Suzuki H, Hashimoto M, Okeya S. Synthesis and crystal structure of $Cs_2(CN_3H_6)_3[(C_2H_5)_2NH_2][W_7O_{22}(O_2)_2]3H_2O$. Eur J Inorg Chem, 2004: 2632-2634.

第8章

八金属氧簇

到目前为止,已经发现的八金属氧簇可分为两大类,一类是人们所熟知的组成为 M_8O_{26} 的八金属氧簇,另一类是最近才发现的组成为 M_8O_{30} 的八金属氧簇。有关人们熟知的八金属氧簇 M_8O_{26} 的报道都集中在钼氧簇形成的化合物上,尽管形成该类八金属氧簇的金属元素种类少,但八钼氧簇却表现出了丰富多彩的异构现象,使得该类化合物阴离子结构种类繁多,目前已经报道的八钼氧簇 M_8O_{26} 有 α、β、γ、δ、ε、ξ、η、θ 等八种异构体。而组成为 M_8O_{30} 的八金属氧簇目前已发现有钼、钨两种构成金属,并且它们以骨架结构的形式存在于阴离子中。下面分别对它们作一介绍。

8.1 八金属氧簇 M_8O_{26}

8.1.1 α 异构体

(1) 合成

向 pH 为 3~4 的钼酸盐溶液中加入四丁基铵阳离子 Bu_4N^+,则可以生成 $(Bu_4N)_4[\alpha\text{-}Mo_8O_{26}]$ 白色沉淀[1,2]。

$$8Na_2MoO_4 + 12HCl + 4(n\text{-}C_4H_9)_4NBr \longrightarrow [(n\text{-}C_4H_9)_4N]_4[\alpha\text{-}Mo_8O_{26}] + 12NaCl + 4NaBr + 6H_2O$$

(2) 谱学表征

八钼氧簇 α 异构体 $[\alpha\text{-}Mo_8O_{26}]^{4-}$ 的红外光谱如图 8-1 所示。

(3) 晶体结构

Fuchs 和 Hartl 首次确定了 α 型八钼氧簇阴离子的结构[1]，随后 Klemperer 和 Day 对该阴离子的结构进行了研究[2,3]。$[\alpha\text{-}Mo_8O_{26}]^{4-}$ 的原子连接方式如图 8-2(a) 所示，多面体连接方式如图 8-2(b) 所示。

由图可知，八钼氧簇阴离子 α 异构体 $[\alpha\text{-}Mo_8O_{26}]^{4-}$ 具有 D_{3d} 对称性，它包含 6 个六配位钼氧八面体和 2 个四配位钼氧四面体，6 个六配位钼氧八面体 $[MoO_6]$ 通过共边相连形成闭合的环，2 个四配位的钼氧四面体 $[MoO_4]$ 分别位于环的两侧通过共顶点与环相连。阴离子中的钼原子按照配位环境的不同可分为两类：一类是位于环上的六配位的钼 [Mo(Ⅰ)]，连接有两个端氧、两个二重桥氧和三个三重桥氧，八面体构型，这类钼原子有 6 个；另一类是位于环的侧面的四配位的钼原子 [Mo(Ⅱ)]，分别连有三个三重桥氧和一个端氧，四面体构型，这类钼原子有 2 个。阴离子中的氧原子分为三类，分别是只与一个钼原子相连的端氧（O_t）、与两个钼原子相连的桥氧（O_b）和与三个钼原子相连的三重桥氧（O_c）。表 8-1 列出了已报道的部分八钼氧簇 α 异构体的键长数据。

图 8-1 $(Bu_4N)_4$ $[\alpha\text{-}Mo_8O_{26}]$ 红外光谱

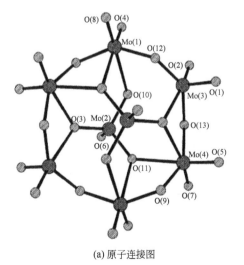

(a) 原子连接图   (b) 多面体

图 8-2 $[\alpha\text{-}Mo_8O_{26}]^{4-}$ 结构

表 8-1　在不同化合物中 $[α\text{-}Mo_8O_{26}]^{4-}$ 的键长数据　　　　单位：Å

化合物阳离子	$Mo_{(I)}\text{—}O_t$	$Mo_{(I)}\text{—}O_b$	$Mo_{(I)}\text{—}O_c$	$Mo_{(II)}\text{—}O_t$	$Mo_{(II)}\text{—}O_c$	文献
$[(n\text{-}C_3H_7)(C_6H_5)_3P]_4^+$	1.696	1.904	2.425	1.708	1.783	[3]
$[(C_2H_5)_4N]_2^+$	1.695	1.910	2.427	1.714	1.779	[4]
$[Cu_2(1,10\text{-}phen)_2(4,4'\text{-}bpy)_2]^{4+}$	1.694	1.909	2.416	1.688	1.786	[5]
$[Co_2(tpyprz)(H_2O)_2]^{4+}$	1.706	1.908	2.386	1.704	1.797	[6]
$[Ni(tpyprz)(H_2O)_2]^{4+}$	1.708	1.908	2.388	1.705	1.790	[6]

(4) 相关化合物

在目前已知的 $[α\text{-}Mo_8O_{26}]^{4-}$ 形成的化合物中，除少数是在水溶液中常规合成的以外[1,2]，大部分都是在水热条件下生成的[5,11]，有时为了向分子内引入金属有机基团，也采用在无氧无水条件下回流的方法合成。由 $(NH_4)_6Mo_7O_{24}\cdot 4H_2O$、$Cu(OOCCH_3)_2\cdot H_2O$、4,4'-联吡啶和 1,10-邻菲啰啉混合溶于水，用 10% 盐酸调 pH 值至 3.56，在 160℃下反应 5d，可得到分子组成为 $[Cu_2(1,10\text{-}phen)_2(4,4'\text{-}bpy)_2]_2[α\text{-}Mo_8O_{26}]\cdot 4H_2O$ 的化合物[5]，其分子结构如图 8-3 所示，由图中可知，$[α\text{-}Mo_8O_{26}]^{4-}$ 和配阳离子之间以静电引力结合。

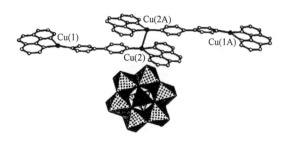

图 8-3　化合物 $[Cu_2(1,10\text{-}phen)_2(4,4'\text{-}bpy)_2]_2[α\text{-}Mo_8O_{26}]\cdot 4H_2O$ 结构

由 MoO_3、4-二吡啶基吡嗪（tpyprz）、$M(CH_3COO)_2\cdot 2H_2O$（M=Co, Ni）和 HF 在 170℃下反应 48h 可得到结构单元为 $[M_2(tpyprz)(H_2O)_2Mo_8O_{26}]\cdot xH_2O$ 的化合物[6]，在该化合物中，α 结构的八钼氧簇阴离子 $[α\text{-}Mo_8O_{26}]^{4-}$ 通过环上的 4 个钼氧八面体上的各一个端基氧原子与过渡金属离子 M^{2+} 配位，构成二维层状结构。由 MoO_3、四-2-吡啶基吡嗪和五水硫酸铜在 160℃下反应 72h，也可以得到单元组成为 $[\{Cu_2(tpyprz)(H_2O)_2\}Mo_8O_{26}]$ 的化合物[7]，该化合物与上述两个化合物同构，也具有二维层状结构（图 8-4）。

由硝酸铜 $Cu(NO_3)_2\cdot 2.5H_2O$、MoO_3 和 1,2-双（吡啶基）乙烯（bpe）在水热 200℃的温度下反应 25.5h，可以得到分子组成为 $[\{Cu(bpe)\}_4(α\text{-}Mo_8O_{26})]\cdot$

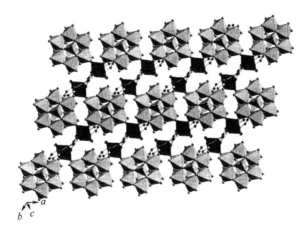

图 8-4 由 $[\alpha\text{-}Mo_8O_{26}]^{4-}$ 与过渡金属离子 M^{2+} 配位构成的二维层状结构

$2H_2O$ 的橙色晶体,该化合物具有二维结构,1,2-双(吡啶基)乙烯铜离子配位形成一维阳离子链 $\{Cu(bpe)\}_n^{h+}$,$[\alpha\text{-}Mo_8O_{26}]^{4-}$ 中位于上、下两侧的 MoO_4 四面体上的每一个端基氧与两个阳离子链上的铜配位,形成图 8-5(a) 所示的结构单元,进而构成二维层状结构 [图 8-5(b)][8]。

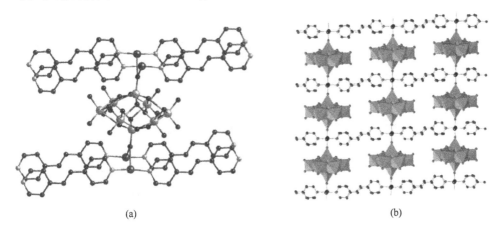

(a)　　　　　　　　　　(b)

图 8-5 化合物结构单元 $[\{Cu(bpe)\}_4^{h+}(\alpha\text{-}Mo_8O_{26})]$ 和二维层状结构

由钼酸钠 $Na_2MoO_4 \cdot H_2O$、五水硫酸铜 ($CuSO_4 \cdot 5H_2O$)、烟酸 (py-3-COOH) 和水在 198℃下反应 4d,可以得到组成为 $[\{Cu(py)_3\}_2\{Cu(py)_2\}_2(\alpha\text{-}Mo_8O_{26})]$ 的化合物[9],在该化合物中,$[\alpha\text{-}Mo_8O_{26}]^{4-}$ 通过六元环上的两个钼氧八面体的端基氧原子与 $[Cu(py)_3]^+$ 配离子中的铜配位,位于环两侧的 MoO_4 四面体的端基氧原子与两个 $[Cu(py)_2]^+$ 配离子中的铜配位,形成如图 8-6 所示的结构,

在晶体中，分子间通过 π—π 堆积相互连接。

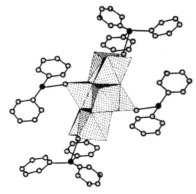

图 8-6 化合物[{Cu(py)$_3$}$_2$ {Cu(py)$_2$}$_2$(α-Mo$_8$O$_{26}$)]结构

由五水硫酸铜（CuSO$_4$·5H$_2$O）、三氧化钼（MoO$_3$）、4,4′-亚丙基联吡啶（dpp）和水在 120℃下反应 48h，可以得到结构单元组成为 [Cu(dpp)]$_2$[Cu$_2$(α-Mo$_8$O$_{26}$)(dpp)$_2$]·2H$_2$O 的化合物[10]。在该化合物中，[α-Mo$_8$O$_{26}$]$^{4-}$ 中六元环两侧的两个钼氧四面体 MoO$_4$ 上的端氧与铜配位，形成链状结构 [图 8-7(a)]，六元环上的两个钼氧八面体 MoO$_6$ 上的端基氧与铜配位，构成二维层状结构 [图 8-7(b)][10]。

由俾斯麦棕-Y 二环己基碳二亚胺

$$\left(H_2N-\bigcirc-N=N-\bigcirc-N=N-\bigcirc_{NH_2}^{H_2N}\right)·HCl$$

和 [n-Bu$_4$N]$_4$[α-Mo$_8$O$_{26}$]在氩气保护下，于无水乙腈中回流 24h，可以得到分子组成为 [n-Bu$_4$N]$_8$[Mo$_6$O$_{19}$]$_2$[α-Mo$_8$O$_{26}$]的化合物[11]。该反应表明，在适当的条件下，[α-Mo$_8$O$_{26}$]$^{4-}$ 可以转化为 Lindqvist 结构的 Mo$_6$O$_{19}^{2-}$。氯化亚钴（CoCl$_2$·6H$_2$O）、钼酸钠（Na$_2$MoO$_4$·2H$_2$O）、2,2′-联吡啶和水混合搅拌，用 HCl 调 pH 到 3.8，然后在 160℃下反应 4d，得到分子组成为 [Co(2,2′-bpy)$_3$]$_4$[α-Mo$_8$O$_{26}$][β-Mo$_8$O$_{26}$]·5H$_2$O 的化合物[12]，该化合物的奇特之处是在一个分

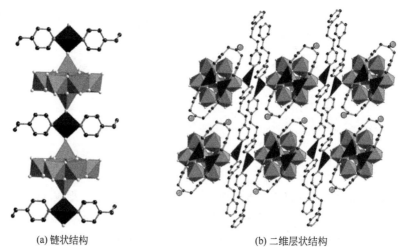

(a) 链状结构　　　　　　　　　　(b) 二维层状结构

图 8-7　[α-Mo$_8$O$_{26}$]$^{4-}$ 中钼氧四面体上端氧与铜配位形成的链状结构和六元环上的两个钼氧八面体 MoO$_6$ 上的端基氧与铜配位形成的二维层次结构

子单元中，α异构体与β异构体共存，表明在反应条件下α异构体与β异构体之间存在平衡。图8-8所示为化合物分子结构。

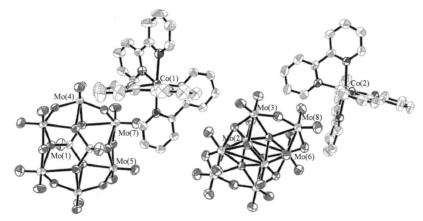

图 8-8 $[Co(2,2'-bpy)_3]_4[\alpha\text{-}Mo_8O_{26}][\beta\text{-}Mo_8O_{26}] \cdot 5H_2O$ 结构

8.1.2 β 异构体

（1）合成

把四烷基卤化铵 $R_4NX(R=CH_3, C_2H_5, n\text{-}C_3H_7; X=Cl, Br)$ 加入酸化过的钼酸钠水溶液中（pH=3~4），可以得到相应的由八钼氧簇阴离子β异构体形成的化合物 $[(CH_3)_4N]_2Na_2[\beta\text{-}Mo_8O_{26}] \cdot 2H_2O$、$[(C_2H_5)_4N]_3Na[\beta\text{-}Mo_8O_{26}]$ 和 $[(n\text{-}C_3H_7)_4N]_2Na_2[\beta\text{-}Mo_8O_{26}] \cdot 2H_2O$[2]。

$$8Na_2MoO_4 + 12HCl + 2(CH_3)_4NCl \longrightarrow [(CH_3)_4N]_2Na_2[\beta\text{-}Mo_8O_{26}] + 14NaCl + 6H_2O$$

四丁基铵盐可以按照下式反应制备[13]：

$$8Na_2MoO_4 + 12HCl + 4(n\text{-}C_4H_9)_4NBr \longrightarrow [(n\text{-}C_4H_9)_4N]_4[Mo_8O_{26}] + 12NaCl + 4NaBr + 6H_2O$$

（2）晶体结构

八钼氧簇β异构体的结构最先由Lindqvist确定[14]，随后有多人对该阴离子的结构进行了研究[2,15-18]。八钼氧簇β异构体 $[\beta\text{-}Mo_8O_{26}]^{4-}$ 的原子连接方式如图8-9(a)所示，多面体结构如图8-9(b)所示。阴离子具有 C_{2h} 对称性，$[\beta\text{-}Mo_8O_{26}]^{4-}$ 由8个畸变的 MoO_6 八面体通过共边和共顶点相连，4个畸变的 MoO_6 八面体相连构成一个四元环，环的中间连接一个共用氧，形成 Mo_4O_{13} 基团，2个 Mo_4O_{13} 基团相连构成 $[\beta\text{-}Mo_8O_{26}]^{4-}$。在阴离子中，氧原子按照连接钼原子的多少，可以分为只与1个钼原子相连的端基氧原子（O_t），阴离子中共14个；与2个钼原子相连的桥氧（O_b），阴离子中共有6个；与3个钼原子相连的三重桥氧（O_c），阴离子中共有4个；与5个钼原子相连的五重桥氧（O_d），阴离子中共有2个。根据键合氧原子的类型，阴离子中的钼原子可以分为三类：连接2个端基氧、3

个二重桥氧和1个五重桥氧的钼原子Mo(Ⅰ),有2个;连接2个端基氧、1个二重桥氧、2个三重桥氧和1个五重桥氧的钼原子Mo(Ⅱ),有4个;连接1个端基氧、1个二重桥氧、2个三重桥氧和2个五重桥氧的钼原子Mo(Ⅲ),有2个。现已报道的晶体数据中,在阴离子中的氧原子不参与配位的情况下,各类键长分别为:Mo—O_t 1.68~1.723Å,Mo—O_b 1.728~2.301Å,Mo—O_c 1.934~2.34Å,Mo—O_d 2.275~2.515Å。

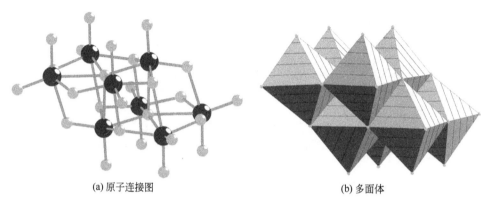

(a) 原子连接图　　　　　　　　　(b) 多面体

图8-9　八钼氧簇β异构体$[\beta\text{-}Mo_8O_{26}]^{4-}$原子连接图和多面体

(3) 相关化合物

由八钼氧簇β异构体$[\beta\text{-}Mo_8O_{26}]^{4-}$形成的化合物是八钼氧簇所有异构体中最多的,为让大家对目前该领域的研究现状有一个比较全面的了解,表8-2列出了由$[\beta\text{-}Mo_8O_{26}]^{4-}$通过静电引力形成的具有独立结构的化合物。

表8-2　$[\beta\text{-}Mo_8O_{26}]^{4-}$形成的具有独立结构的化合物

化合物	反应条件	文献
$[(CH_3)_2NH_2]_4[Mo_8O_{26}] \cdot 2C_3H_7NO$	MoO_3,DMF,回流7h	[19]
$[C_6H_{16}N]_3(H_3O)[Mo_8O_{26}] \cdot 2H_2O$	$[\{i\text{-}(CH_3)_2CH\}NH_2]Mo_2O_7 \cdot 2H_2O$水溶液,黑暗中放置两周	[20,21]
$[Htmpd]_2[tmpd]_2[Mo_8O_{26}]$	N,N,N',N'-四甲基对苯二胺,$[NBu_4][Mo_6O_{19}]$,乙腈,室温	[22]
$[(CH_3)_2NH_2][Bu_4N]_2[Mo_8O_{26}]$	$[Bu_4N][\alpha\text{-}Mo_8O_{26}]$、三乙胺,DMF中回流	[23]
$[TTF]_7Mo_8O_{26}$	TTF,$[\beta\text{-}Mo_8O_{26}]^{4-}$,乙腈中电化学合成	[24]
$[NH_3^iPr]_4[Mo_8O_{26}]$	MoO_3、异丙胺,水溶液回流	[25]
$[Et_3NH]_4[Mo_8O_{26}]$	MoO_3、三乙胺,水溶液回流或钼酸铵、三乙胺、盐酸,水溶液回流	[26]
$[Et_3NH]_3[NaMo_8O_{26}]$	$[Et_3NH]_4Mo_8O_{26}$、NaCl,水溶液回流或钼酸钠、三乙胺、盐酸,水溶液回流	[26]

续表

化合物	反应条件	文献
$[C_5H_{10}NH_2][Mo_8O_{26}] \cdot 4H_2O$	$Na_2MoO_4 \cdot 2H_2O$,哌啶,低 pH 值水溶液回流	[27]
$[Et_3NH]_2[Mg(H_2O)_6Mo_8O_{26}] \cdot 2H_2O$	$[Et_3NH]_4Mo_8O_{26}$,$MgCl_2$,水溶液回流	[28]
$[C_4H_{10}NO]_4[Mo_8O_{26}] \cdot 4H_2O$	Mo_2Cl_2,吗啉甲醇溶液,pH=6	[29]
$[C_6H_{15}NH]_4[Mo_8O_{26}] \cdot 2H_2O$	三乙胺、MoO_3、水溶液回流	[30]
$[Bu_4NH]_3[Mo_8O_{26}]$	$[Bu_4N]_4[\alpha\text{-}Mo_8O_{26}]$,对羟基硫代苯酚、$CuCl_2$ 甲醇溶液,室温	[31]
$[C_3H_5N]_4[Mo_8O_{26}]$	$[C_3H_5N]_4[(C_3H_4N)_2(\nu\text{-}Mo_8O_{26})]$、$Cu(NO_3)_2 \cdot 3H_2O$,水溶液回流	[32]
$[Ni(phen)_3]_2[Mo_8O_{26}] \cdot 2H_2O$	MoO_3、H_2MoO_4、$Ni(Ac)_2 \cdot 6H_2O$、1,10-邻菲啰啉,170℃反应 3d	[33]
$[Cu_4(2\text{-}pzc)_4(H_2O)_8][Mo_8O_{26}] \cdot 2H_2O$	$Cu(NO_3)_2 \cdot 3H_2O$、$(NH_4)_6Mo_7O_{24} \cdot 4H_2O$、2-吡嗪甲酸	[34]
$[C_{10}H_{10}N_2]_2[Mo_8O_{26}]$	H_2MoO_4、$NiCl_2 \cdot 2H_2O$、4,4′-联吡啶、65%(质量分数)HNO_3,443K 反应 6d	[35]
$K_4[Co^{II}(H_2O)_6][Co^{III}(C_{10}H_{12}N_2O_8)]_2[Mo_8O_{26}] \cdot 6H_2O$	$K_3[CoMo_6O_{18}(OH)_6] \cdot nH_2O$、$Na_2H_2EDTA$、$CoCl_2$ 水溶液	[36]
$(C_7H_{11}N_2O)_4[Mo_8O_{26}]$	$Na_2MoO_4 \cdot 2H_2O$、N-2-羟乙基-4-氨基吡啶、$Zn(CH_3COO)_2 \cdot 4H_2O$、$HNO_3$,443K 反应 2d	[37]
$(BEDT\text{-}TTF)_6[Mo_8O_{26}] \cdot 3DMF$	BEDT-TTF、$[Bu_4N]_2[Mo_8O_{26}]$,DMF 溶液,电化学合成	[38]
$(C_{10}H_{18}N)_4Mo_8O_{26} \cdot 6(CH_3)_2SO$	$Na_2MoO_4 \cdot 2H_2O$、金刚烷胺、HCl,DMSO 重结晶	[39]
$[Co(phen)_3]_2[Mo_8O_{26}] \cdot 2.5H_2O$	$CoCl_2 \cdot 6H_2O$、$(NH_4)_6Mo_7O_{24} \cdot 4H_2O$、$SeO_2$、邻菲啰啉,pH=5~6,200℃反应 6d	[40]
$[Fe(2,2'\text{-}bpy)_3]_2[Mo_8O_{26}] \cdot 6H_2O$	MoO_3、2,2′-联吡啶、$FeSO_4$,pH=6.0,180℃反应 3d	[41]
$[Co(H_2O)_6][(CH_3)_4N][Mo_8O_{26}]$ 4 甲胺为 2 个	$Na_2MoO_4 \cdot 2H_2O$、MoO_3、$NH_2OH \cdot HCl$、$(CH_3)_4NCl$、$Co(CH_3COO)_2$,433K 反应 2d	[42]
$[(CH_3)_4N]_4[Mo_8O_{26}] \cdot 2H_2O$	$[(CH_3)_4N]OH$、MoO_3,pH=7,373K 晶化	[43]
$[Mn(2,2'\text{-}bpy)_2]_2[Mo_8O_{26}] \cdot 4H_2O$ 3 个联吡啶	$MnCl_2 \cdot 2H_2O$、2,2′-联吡啶、$(NH_4)_6Mo_7O_{24} \cdot 4H_2O$、$K_3[Fe(CN)_6]$,pH=3,170℃,5d	[44]
$[4\text{-}MePyH]_4[Mo_8O_{26}]$	4-MePy 缓慢扩散到 $MoO_2(Py\text{-}S)_2$ 乙腈溶液中	[45]
$[TTF]_7Mo_8O_{26}$	TTF、$[\alpha\text{-}Mo_8O_{26}]^{4-}$ 或 $[\beta\text{-}Mo_8O_{26}]^{4-}$,乙腈溶液	[46]
$[C_{15}H_{17}N_4]_4[Mo_8O_{26}]$	$(NH_4)_6Mo_7O_{24} \cdot 4H_2O$,氯化中性红,180℃,3d	[47]

续表

化合物	反应条件	文献
$(C_5H_9N_2)(H_3O)[Mo_8O_{26}]\cdot 2C_2H_6O$ 阳离子均是2	$MoO_2(acac)_2$乙醇溶液,3,5-二甲基吡咯水溶液,回流	[48]
$(NH_4)_4[Mo_8O_{26}]\cdot 2C_8H_8N_2$	$CuCl_2\cdot 2H_2O,(NH_4)_6Mo_7O_{24}\cdot 4H_2O,MoO_3$、2-甲基苯并咪唑,443K,3d	[49]
$(C_6H_{16}N_2)_2[Mo_8O_{26}]$	$MoO_3,H_2SO_4,2,5$-二甲基吡嗪,453K,1d	[50]
$[C_{12}H_{11}N_3]_2[Mo_8O_{26}]$	$MoO_3,ZnSO_4\cdot 7H_2O,2$-(4-吡啶基)苯并咪唑,453K,3d	[51]

八钼氧簇β异构体$[\beta\text{-}Mo_8O_{26}]^{4-}$除了形成分离结构的化合物以外,还可以通过端基氧或桥氧与金属原子配位,形成具有新颖结构的化合物。$[\beta\text{-}Mo_8O_{26}]^{4-}$可以通过端基氧与碱金属钠、钾离子配位,形成具有链状、层状结构的化合物[52-55]。图8-10为化合物$(4,4'\text{-}Hbpy)_2[K_2Mo_8O_{26}]$中阴离子的结构单元和阴离子的层状结构。

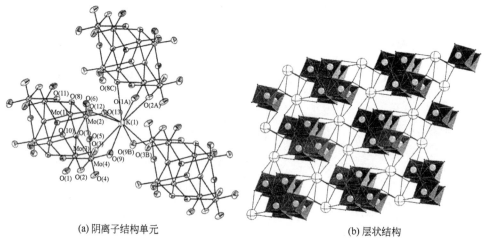

(a) 阴离子结构单元　　　　　　　　(b) 层状结构

图8-10　化合物$(4,4'\text{-}Hbpy)_2[K_2Mo_8O_{26}]$中阴离子结构单元和层状结构

$[\beta\text{-}Mo_8O_{26}]^{4-}$与碱土金属钡配位,在有机配体DMF的参与下,可生成组成为$[Ba(DMF)_2(H_2O)_2]_2[\beta\text{-}Mo_8O_{26}]\cdot 2DMF$的结构单元 (图8-11)[56]。在结构单元中,$[\beta\text{-}Mo_8O_{26}]^{4-}$中的八个端氧分别与两个$Ba^{2+}$配位,与$Ba^{2+}$配位的水分子和DMF分子作为桥联配体,与另一个结构单元中的Ba^{2+}配位,形成一维链状结构,相邻链之间通过$[\beta\text{-}Mo_8O_{26}]^{4-}$中的两个端氧,再和$Ba^{2+}$配位,形成二维层状结构 (图8-12)。

$[\beta\text{-}Mo_8O_{26}]^{4-}$与过渡金属的配位研究主要集中在铜和银元素上。由0.1g $Cu(CH_3COO)_2\cdot H_2O$,$0.242gNa_2MoO_4\cdot 2H_2O$和0.1g喹喔啉在18mL水中混合,用浓盐酸调pH值为4.0,在170℃下反应5d,可得到结构单元为$[Cu_2(C_8H_6N_2)_2(C_7H_6N_2)]_2[Mo_8O_{26}]$的配位聚合物[57]。图8-13为该配位聚合

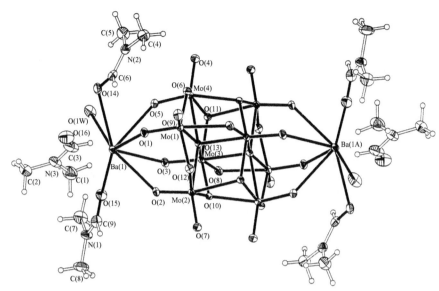

图 8-11 化合物 $[Ba(DMF)_2(H_2O)_2]_2[\beta\text{-}Mo_8O_{26}]\cdot 2DMF$ 结构单元

物结构中阴离子的配位连接方式,每个阴离子通过八个端基氧原子与四个铜原子配位,每一个铜原子同时与阴离子中的两个端基氧、喹喔啉分子中的一个氮和苯并咪唑中的一个氮原子配位,苯并咪唑中的另一个氮与和另一个阴离子相连的铜原子配位,形成一维链状结构(图 8-14)。值得注意的是反应原料中并没有加入苯并咪唑,化合物中的苯并咪唑配体可能来源于喹喔啉的氧化[57]。

图 8-12 化合物 $[Ba(DMF)_2(H_2O)_2]_2$ $[\beta\text{-}Mo_8O_{26}]\cdot 2DMF$ 形成的二维层状结构

由 $MoO_3\cdot H_2O$、$(NH_4)_6Mo_7O_{24}\cdot 4H_2O$、$CuSO_4\cdot 5H_2O$、2,2'-联吡啶和水在 170℃下反应 2d,可以得到分子组成为 $[\{Cu(2,2'\text{-}bpy)\}_2(\beta\text{-}Mo_8O_{26})]$ 的

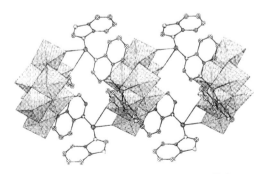

图 8-13 化合物 $[Cu_2(C_8H_6N_2)_2(C_7H_6N_2)]_2[Mo_8O_{26}]$ 中阴离子的配位连接方式

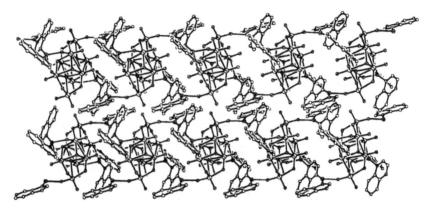

图 8-14 化合物$[Cu_2(C_8H_6N_2)_2(C_7H_6N_2)]_2[Mo_8O_{26}]$
构成的一维链状结构

化合物[9]，在该化合物中，$[\beta\text{-}Mo_8O_{26}]^{4-}$通过两个二重桥氧原子与$[Cu(2,2'\text{-}bpy)_2]^{2+}$配离子相连，形成具有担载结构的中性分子（图 8-15）。

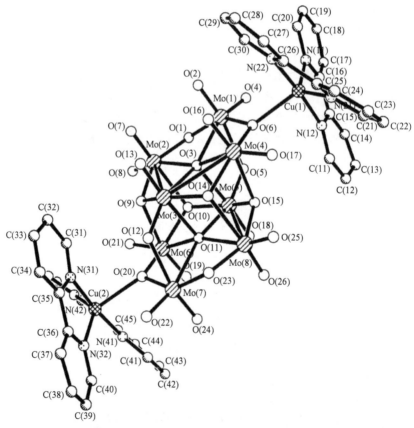

图 8-15 化合物$[\{Cu(2,2'\text{-}bpy)\}_2(\beta\text{-}Mo_8O_{26})]$形成的担载结构

MoO_3、H_2MoO_4、$Cu(Ac)_2 \cdot H_2O$、1,10-邻菲啰啉和水在170℃下反应3d，得到分子组成为$[Cu(phen)_2]_2[\{Cu(phen)\}_2Mo_8O_{26}] \cdot H_2O$的化合物[58]，在该化合物中，$[\beta-Mo_8O_{26}]^{4-}$中的四个氧原子与两个一价的$Cu^+$配位，形成具有担载结构的$[\{Cu(phen)\}_2Mo_8O_{26}]^{2-}$，该阴离子通过静电作用力与配阳离子$[Cu(phen)_2]^+$结合，形成化合物。图8-16表明了$[\{Cu(phen)\}_2Mo_8O_{26}]^{2-}$的原子连接方式。

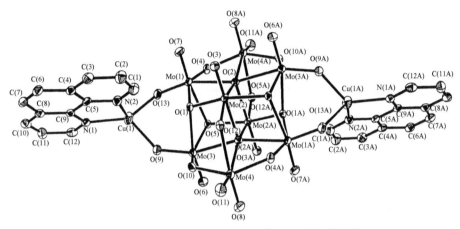

图8-16　$[\{Cu(phen)\}_2Mo_8O_{26}]^{2-}$形成的担载结构

MoO_3、CuO、吡嗪-2-甲酸和水在130℃下反应72h得到分子结构单元为$[Cu_5(pzca)_6(H_2O)_4][Mo_8O_{26}]$的配位聚合物[59]，在该结构单元中，$[\beta-Mo_8O_{26}]^{4-}$通过两个端基氧原子与两个桥联基团中的$Cu^{2+}$配位，两个桥联基团中的$Cu^{2+}$再与另外两个$[\beta-Mo_8O_{26}]^{4-}$相连，形成一维链状结构（图8-17）。

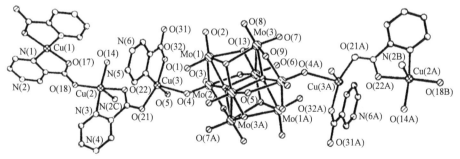

图8-17　配位聚合物$[Cu_5(pzca)_6(H_2O)_4][Mo_8O_{26}]$的结构单元

170℃水热条件下，MoO_3、$CuCl_2 \cdot 2H_2O$、$Na_2MoO_4 \cdot H_2O$和苯并咪唑反应72h，得到分子组成为$[Cu(bIz)_2]_2[\{Cu(bIz)_2\}_2Mo_8O_{26}]$的化合物[60]。化合物在生成的过程中，二价铜被还原为一价铜。化合物阴离子$[\{Cu(bIz)_2\}_2Mo_8O_{26}]^{2-}$

通过静电引力和配阳离子 $Cu(bIz)_2^+$ 结合。阴离子中，$[\beta\text{-}Mo_8O_{26}]^{4-}$ 通过两个端基氧原子与两个 $Cu(bIz)^+$ 配离子结合 (图 8-18)。

$Na_2MoO_4 \cdot 2H_2O$、2,2'-联吡啶和 $NiCl_2 \cdot 6H_2O$，用 HAc 调 pH=2.6，室温放置可缓慢析出分子组成为 $[Ni(H_2O)(2,2'\text{-}bpy)_2]_2[Mo_8O_{26}] \cdot 4H_2O \cdot 2CH_3COOH$ 的化合物[61]。在该化合物分子中，$[\beta\text{-}Mo_8O_{26}]^{4-}$ 通过两个端基氧原子与两个 $[Ni(H_2O)(2,2'\text{-}bpy)_2]^{2+}$ 相连，形成担载结构的中性分子 (图 8-19)。

图 8-18　化合物 $[Cu(bIz)]_2[\{Cu(bIz)_2\}_2Mo_8O_{26}]$ 结构

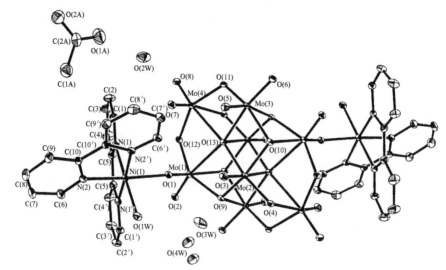

图 8-19　化合物 $[Ni(H_2O)(2,2'\text{-}bpy)_2]_2[Mo_8O_{26}] \cdot 4H_2O \cdot 2CH_3COOH$ 结构

MoO_3、$CoSO_4 \cdot 7H_2O$、2,2'-联吡啶和水，用 1,2-丙二胺调 pH 到 6.0，160℃下反应 6d，得到结构单元组成为 $[Co^{II}(2,2'\text{-}bpy)_2]_2[Mo_8O_{26}]$ 的配位聚合物[62]。在该化合物中，$[\beta\text{-}Mo_8O_{26}]^{4-}$ 中的 4 个端基氧与 4 个 $Co(2,2'\text{-}bpy)_2^{2+}$ 配离子配位，每一个 $Co(2,2'\text{-}bpy)^{2+}$ 配离子与两个 $[\beta\text{-}Mo_8O_{26}]^{4-}$ 相连，构成一维链状结构 [图 8-20(a)、(b)]。

$[Bu_4N]_2[Mo_6O_{19}]$ 的 DMF 溶液和 $AgNO_3$ 的甲醇溶液室温搅拌放置，可得到结构单元组成为 $\{[Bu_4N]_2[Ag_2Mo_8O_{26}]\}$ 的化合物[63]。在该化合物中，每一个 $[\beta\text{-}Mo_8O_{26}]^{4-}$ 上的八个端基氧原子两两一组分别与四个 Ag^+ 配位[图 8-21(a)]，每一个银离子与两个 $[\beta\text{-}Mo_8O_{26}]^{4-}$ 相连，形成一维链状结构[图 8-21(b)]。

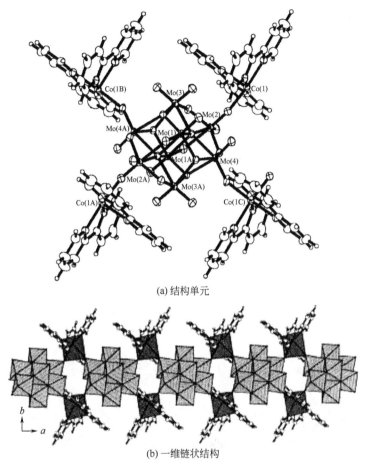

(a) 结构单元

(b) 一维链状结构

图 8-20　化合物 $[Co^{II}(2,2'-bpy)_2]_2[Mo_8O_{26}]$ 结构单元和一维链状结构

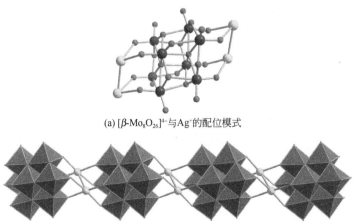

(a) $[\beta\text{-}Mo_8O_{26}]^{4-}$ 与 Ag^+ 的配位模式

(b) 化合物 $[Bu_4N]_2[Ag_2Mo_8O_{26}]$ 阴离子一维链状结构

图 8-21　$[\beta\text{-}Mo_8O_{26}]^{4-}$ 与 Ag^+ 的配位模式及化合物 $[Bu_4N]_2[Ag_2Mo_8O_{26}]$ 阴离子一维链状结构

由 $(NH_4)_6Mo_7O_{24} \cdot 2H_2O$、$AgNO_3$、2,2'-联吡啶吩嗪和水在 160℃ 温度下反应 5d，得到结构单元组成为 $(H_3O)[Ag_3(2,2'-bpy)_2(phnz)_2(\beta-Mo_8O_{26})]$ 的化合物[64]，在该化合物中，每一个 $[\beta-Mo_8O_{26}]^{4-}$ 通过两个端基氧和两个二重桥氧分别与四个 Ag^+ 配位，四个银离子分为两类，一类与一个二重桥氧、联吡啶中的两个氮原子和吩嗪中的一个氮原子配位；另一类和两个吩嗪中的各一个氮原子和两个阴离子中各一个端基氧原子配位，形成一维链状结构[图 8-22(a)、(b)]。

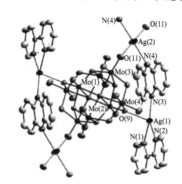

(a) 阴离子配位模式

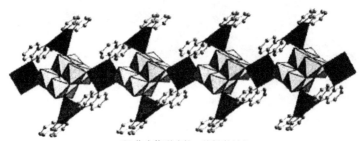

(b) 化合物形成的一维链状结构

图 8-22　化合物 $(H_3O)[Ag_3(2,2'-bpy)_2(phnz)_2(\beta-Mo_8O_{26})]$
中阴离子配位模式及化合物形成的一维链状结构

Cronin 等以 $[Bu_4N]_2[Mo_6O_{19}]$ 和氟化银或硝酸银为原料，通过改变溶剂条件，制得了组成为 $[Bu_4N]_2[Ag_2Mo_8O_{26}]$、$[Bu_4N]_2[Ag_2Mo_8O_{26}(CH_3CN)_2]$ 和 $[Bu_4N]_2[Ag_2(Mo_8O_{26})\{(CH_3)_2SO\}_2]$ 的化合物[65]。化合物 $[Bu_4N]_2[Ag_2Mo_8O_{26}]$ 的结构与稍早报道的很像[63]，但在该化合物中，阴离子上有两个端基氧原子同时与两个 Ag^+ 配位，银具有平面正方形构型（图 8-23）。

在化合物 $[Bu_4N]_2[Ag_2Mo_8O_{26}(CH_3CN)_2]$ 中，阴离子同样具有链状结构，但由于 CH_3CN 分子参与了配位，链中的桥联银离子为六配位八面体构型（图 8-24）。化合物 $[Bu_4N]_2[Ag_2Mo_8O_{26}\{(CH_3)_2SO\}_2]$ 的结构与前两者有明显的差别，首先是化合物中存在沿两个方向伸展的链，这两个方向上链的夹角是 86°（图 8-25），另外，每一条链中 Ag^+ 与 $[\beta-Mo_8O_{26}]^{4-}$ 的连接方式也与前两者不同，而是每个银

第8章 八金属氧簇   121

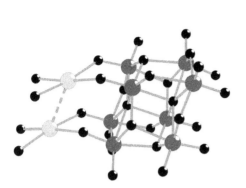

图 8-23 化合物[Bu$_4$N]$_2$[Ag$_2$Mo$_8$O$_{26}$]
中阴离子与 Ag$^+$ 的配位环境

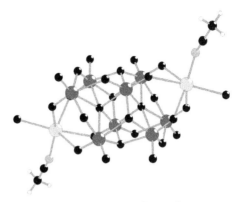

图 8-24 化合物[Bu$_4$N]$_2$
[Ag$_2$Mo$_8$O$_{26}$(CH$_3$CN)$_2$]中
阴离子与 Ag$^+$ 的配位环境

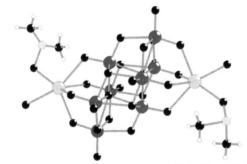

(a) 化合物[Bu$_4$N]$_2$[Ag$_2$Mo$_8$O$_{26}${(CH$_3$)$_2$SO}$_2$]中阴离子和Ag$^+$的配位方式

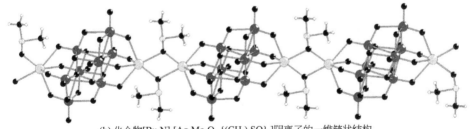

(b) 化合物[Bu$_4$N]$_2$[Ag$_2$Mo$_8$O$_{26}${(CH$_3$)$_2$SO}$_2$]阴离子的一维链状结构

图 8-25 化合物[Bu$_4$N]$_2$[Ag$_2$Mo$_8$O$_{26}${(CH$_3$)$_2$SO}$_2$]中阴离子和
Ag$^+$ 的配位方式及阴离子的一维链状结构

离子与 [β-Mo$_8$O$_{26}$]$^{4-}$ 中的四个端基氧原子配位，两个银离子之间通过两个DMSO 分子中的氧原子桥联形成一维链状结构（图 8-26）。其中的银离子为六配位，畸变三棱柱构型。

在反应体系中用(Pr$_4$N)$_2$[Mo$_6$O$_{19}$]代替[Bu$_4$N]$_6$[Mo$_6$O$_{19}$]，在 DMF 中反应，得到组成为(C$_3$H$_8$NO)[Ag$_3$(Mo$_8$O$_{26}$)(C$_3$H$_7$NO)$_4$]的结构单元。有意思

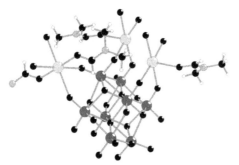

图 8-26 化合物 (C_3H_8NO) $[Ag_3Mo_8O_{26}(C_3H_7NO)_4]$ 中阴离子与银离子的配位环境

是反应原料中的 Pr_4N^+ 没有进入到反应产物中,而是被质子化的 $DMFH^+$ 代替。在阴离子中,$[\beta\text{-}Mo_8O_{26}]^{4-}$ 首先采取与前两种相同的方式,每个阴离子通过八个端基氧和四个银离子配位,每个银离子同时与两个 $[\beta\text{-}Mo_8O_{26}]^{4-}$ 中的各两个端基氧配位,形成一维链状结构,同时银离子还与一个 DMF 中的氧原子配位,形成四方锥结构。一维链中的 $[\beta\text{-}Mo_8O_{26}]^{4-}$ 再通过四个端基氧原子,与位于链两侧的两个银配离子 $Ag(DMF)_2^+$ 相连,构成二维层状结构(图 8-27)。位于链两侧的桥联银配离子 $Ag(DMF)_2^+$ 除与两个 DMF 分子中的氧原子配位外,还与两条相邻链中的两个 $[\beta\text{-}Mo_8O_{26}]^{4-}$ 中的各两个端基氧相连,形成六配位畸变八面体构型。以 $(Ph_4P)_2[Mo_6O_{19}]$ 代替 $(Bu_4N)_2[Mo_6O_{19}]$ 作原料,在 DMSO 中反应,得到分子组成为 $[Ph_4P]_2[Ag_2Mo_8O_{26}\{(CH_3)_2SO\}_4]$ 的化合物,该化合物具有分立结构,阴离子结构如图 8-28 所示。$[\beta\text{-}Mo_8O_{26}]^{4-}$ 通过两组位于四

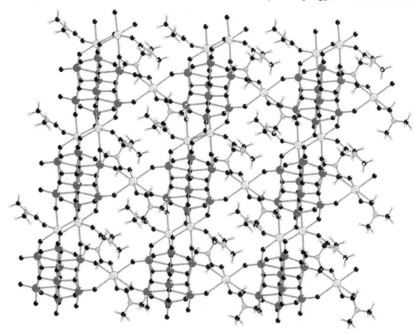

图 8-27 化合物 $(C_3H_8NO)[Ag_3(Mo_8O_{26})(C_3H_7NO)_4]$ 阴离子构成的二维层状结构

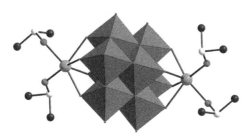

图 8-28　[Ph$_4$P]$_2$[Ag$_2$Mo$_8$O$_{26}${(CH$_3$)$_2$SO}$_4$]阴离子结构

方平面上的端基氧与两个银离子配位,每个银离子同时再与两个 DMSO 中的氧原子配位,形成六配位三棱柱结构。

由文献 [65] 对 5 个不同结构化合物的研究可以看出,在该体系中,阳离子在构成化合物的空间结构中起着非常重要的作用。

用 [Bu$_4$N]$_4$[α-Mo$_8$O$_{26}$] 和 Ln(NO$_3$)$_3$·6H$_2$O(Ln=Y,Ce,Pr,Nd,Gd 和 Yb) 按摩尔比 2∶1 反应,生成分子组成为(Bu$_4$N)$_5$[Ln(Mo$_8$O$_{26}$)$_2$]的化合物[66]。

$$2[Mo_8O_{26}]^{4-} + Ln(NO_3)_3 \longrightarrow [Ln(Mo_8O_{26})_2]^{5-} + 3NO_3^-$$

在该化合物中,[Ln(Mo$_8$O$_{26}$)$_2$]$^{5-}$ 由两个 [β-Mo$_8$O$_{26}$]$^{4-}$ 分别通过四个位于四方平面上的端基氧与稀土离子配位,形成一种夹心结构,图 8-29 为 [Ln(Mo$_8$O$_{26}$)$_2$]$^{5-}$ 的原子连接图,其中稀土离子为八配位四方反棱柱构型。改变反应物摩尔比为[Bu$_4$N]$_4$[α-Mo$_8$O$_{26}$]∶Ln(NO$_3$)$_3$·6H$_2$O=1∶2,则体系有如下反应:

$$[Mo_8O_{26}]^{4-} + 2Ln(NO_3)_3 \longrightarrow [\{Ln(NO_3)_3\}_2(Mo_8O_{26})]^{4-}$$

两种不同结构类型的化合物[Bu$_4$N]$_4$[{Ln(NO$_3$)$_3$}$_2$(Mo$_8$O$_{26}$)]和(Bu$_4$N)$_5$[Ln(Mo$_8$O$_{26}$)$_2$]的生成表明了反应物的比例对产物结构类型的影响。在[{Ln(NO$_3$)$_3$}$_2$(Mo$_8$O$_{26}$)]$^{4-}$ 中,[β-Mo$_8$O$_{26}$]$^{4-}$ 通过两组位于四方形平面上的各四个端基氧原子分别与两个稀土离子配位,每个稀土离子在与 [β-Mo$_8$O$_{26}$]$^{4-}$ 中的四个氧原子配位的同时,还与六个来自三个硝酸根的氧原子配位,形成十配位构型(图 8-30)。

仲钼酸铵 (NH$_4$)$_6$Mo$_7$O$_{24}$·4H$_2$O 和氯化钆 GdCl$_3$ 在 pH=1.76 的水和 DMF 溶液中室温反应,得到分子组成为[NH$_4$]$_2$[{Gd(DMF)$_7$}$_2$(β-Mo$_8$O$_{26}$)][β-Mo$_8$O$_{26}$]的化合物[67]。在该化合物中,除了 [NH$_4$]$^+$ 和 [β-Mo$_8$O$_{26}$]$^{4-}$ 外,还存在一个 [{Gd(DMF)$_7$}$_2$(β-Mo$_8$O$_{26}$)]$^{2+}$,在该阳离子中,[β-Mo$_8$O$_{26}$]$^{4-}$ 通过两个端氧原子与两个 Gd(DMF)$_7^{3+}$ 相连,形成具有担载结构的配阳离子 (图 8-31)。

由 (NH$_4$)$_3$Mo$_7$O$_{24}$ 和 LaCl$_3$ 在乙腈、DMF 和水的混合溶剂中,在 pH=2.32

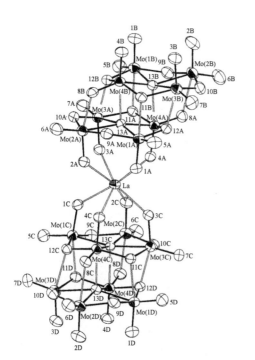

图 8-29 化合物 $(Bu_4N)_5[Ln(Mo_8O_{26})_2]$ 中阴离子原子连接图

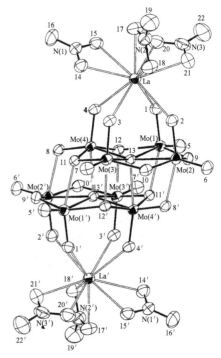

图 8-30 $[\{Ln(NO_3)_3\}_2(Mo_8O_{26})]^{4-}$ 原子连接图

图 8-31 $[\{Gd(DMF)_7\}_2(\beta\text{-}Mo_8O_{26})]^{2+}$ 结构

的酸度下室温反应，得到分子组成为 $[NH_4][La(DMF)_7(\beta\text{-}Mo_8O_{26})]$ 的化合物[67]。在该化合物中，阴离子由 $[\beta\text{-}Mo_8O_{26}]^{4-}$ 通过两个端基氧原子与一个 $La(DMF)_7^{3+}$ 结合，形成具有担载结构的配阴离子（图 8-32）。

$(Bu_4N)_2[Mo_6O_{19}]$ 和钼酸钠的 DMF 溶液与 $La(NO_3)_3 \cdot 3H_2O$ 的乙醇溶液在 70℃下反应，可得到结构单元组成为 $NaLaMo_8O_{26}(C_3H_7NO)_7$ 的链状配位聚合物[68]。在该化合物中，$[\beta\text{-}Mo_8O_{26}]^{4-}$ 中的两组四方平面氧原子分别与两个钠离子相连，钠离子再与另一个相邻的 $[\beta\text{-}Mo_8O_{26}]^{4-}$ 中的一组四方平面氧原子相连，形成一维链状结构。另外，阴离子还通过两个端基氧原子与配离子

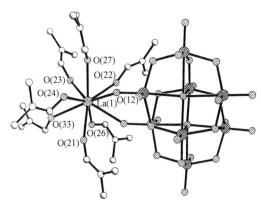

图 8-32 化合物 $[NH_4][La(DMF)_6(\beta\text{-}Mo_8O_{26})]$ 中阴离子结构

$[La(DMF)_7]^{3+}$ 相连（图 8-33），形成包含担载结构单元的一维链状化合物。在该化合物中，La^{3+} 为九配位。

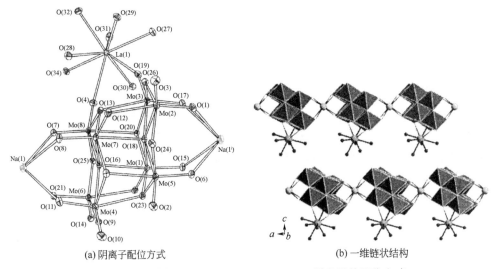

(a) 阴离子配位方式　　　　　　(b) 一维链状结构

图 8-33　化合物 $NaLaMo_8O_{26}(C_3H_7NO)_7$ 阴离子的配位方式和一维链状结构

8.1.3　γ 异构体

(1) 合成

有关八钼氧簇 γ 异构体的报道目前还较少。具有典型独立结构的 $[\gamma\text{-}Mo_8O_{26}]^{4-}$ 异构体目前仅有一例报道[69]。其合成方法如下，用 1.0mol/L 的盐酸滴加到每升含有 0.4mol$Na_2Mo_4\cdot 2H_2O$ 和 0.04mol$[Me_3(CH_2)_6NMe_3]Cl_2$ 的水溶液中，酸滴加会有白色沉淀生成，搅拌下沉淀迅速溶解，在 pH 为 6 左右时，生成不溶的微晶，放置可得到质量良好的晶体。

（2）结构

八钼氧簇 γ 异构体 $[\gamma\text{-}Mo_8O_{26}]^{4-}$ 的原子连接图和多面体如图 8-34 所示，阴离子具有 C_i 对称性。

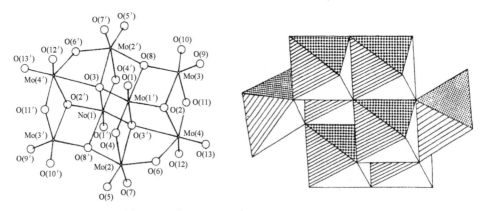

图 8-34　$[\gamma\text{-}Mo_8O_{26}]^{4-}$ 原子连接图和多面体

阴离子由两个 MoO_5 四方锥和六个 MoO_6 八面体通过共边和共顶点相连而成。根据连接钼原子个数的不同，阴离子中的氧原子可分为：端基氧原子（O_t），共有 14 个；二重桥氧原子（O_b），共 6 个；三重桥氧原子（O_c），共 4 个；四重桥氧原子（O_d），共 2 个。按照键合氧原子的种类，八个钼原子可分为三类：一类 Mo_I，连接 2 个端基氧原子 O_t、1 个二重桥氧原子 O_b 和 2 个三重桥氧原子 O_c，Mo_I 原子具有四方锥构型，有 2 个；第二类钼原子 Mo_{II}，连接 2 个端基氧原子 O_t、2 个二重桥氧原子 O_b、1 个三重桥氧原子 O_c 和 1 个四重桥氧原子 O_d，Mo_{II} 原子为八面体构型，有 4 个；第三类 Mo_{III}，连接 1 个端基氧原子 O_t、1 个二重桥氧原子 O_b、2 个三重桥氧原子 O_c 和 2 个四重桥氧原子 O_d，Mo_{III} 原子为八面体构型，有 2 个。

（3）相关化合物

由 $CuSO_4 \cdot 5H_2O$、咪唑、$MoO_3 \cdot H_2O$ 和水在 170℃ 下反应 7d，可以得到分子组成为 $[Cu(imi)_2]_4[\gamma\text{-}Mo_8O_{26}]$ 的化合物。该化合物分子为具有担载结构的中性分子[70]，化合物中的铜为 +1 价，$[\gamma\text{-}Mo_8O_{26}]^{4-}$ 通过六个端基氧原子与四个 $[Cu(imi)_2]^+$ 配离子相连，其中两个配离子 $[Cu(imi)_2]^+$ 只与一个端基氧原子配位，另两个配离子 $[Cu(imi)_2]^+$ 分别与两个钼原子上的各一个氧原子配位（图 8-35）。

尽管具有分立结构的八钼氧簇 γ 异构体阴离子 $[\gamma\text{-}Mo_8O_{26}]^{4-}$ 在 1991 年才有报道[69]，但 $[\gamma\text{-}Mo_8O_{26}]^{4-}$ 的衍生结构化合物此前已经制备出来。Isobe 等 1978 年报道的组成为 $(C_3H_{10}N)_6[H_2Mo_8O_{28}] \cdot 2H_2O$ 的化合物中[71]，

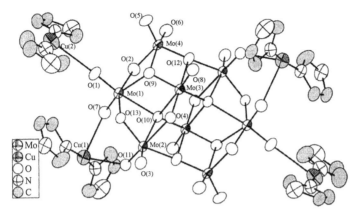

图 8-35　化合物 [Cu(imi)$_2$]$_4$[γ-Mo$_8$O$_{26}$] 结构

[H$_2$Mo$_8$O$_{28}$]$^{6-}$ 中去掉两个 OH$^-$ 基团，剩余部分即为 [γ-Mo$_8$O$_{26}$]$^{4-}$。1979 年，Adams 用甲酸和 (NH$_4$)$_6$[Mo$_7$O$_{24}$] 水溶液反应，制得了分子组成为 (NH$_4$)$_6$[(HCO)$_2$Mo$_8$O$_{28}$]·2H$_2$O 的化合物[72]，化合物阴离子结构如图 8-36 所示。

该化合物的阴离子化学组成可以写作 [γ-Mo$_8$O$_{26}$(HCOO)$_2$]$^{4-}$，为甲酸根衍生物，阴离子中的 2 个甲酸根占据了 [γ-Mo$_8$O$_{26}$]$^{4-}$ 中两个五配位四方锥构型钼原子的底部，钼原子恢复为八面体构型。MoO$_3$·2H$_2$O 与过量的无水乙醇在 4Å 分子筛存在下反应，过滤除去主要反应产物白色固体 Mo$_2$O$_5$(OCH$_3$)$_2$，缓慢蒸发溶液，可得到分子组成为 Na$_4$[Mo$_8$O$_{24}$(OCH$_3$)$_4$]·8CH$_3$OH 的化合物[73]，化合物阴离子结构如图 8-37 所示，阴离子中有四个甲氧基，两个甲氧基作为桥联基团，连接两个钼原子，另外两个甲氧基分别与

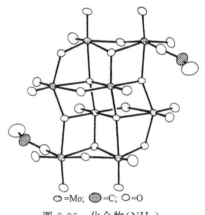

图 8-36　化合物 (NH$_4$)$_6$[(HCO)$_2$Mo$_8$O$_{28}$]·2H$_2$O 中阴离子结构

一个钼原子相连。与一个钼原子相连的甲氧基，占据了 γ 异构体中五配位四方锥构型钼原子构成的四方锥的底部。

1984 年，Mccarron 报道，用 MoO$_3$·2H$_2$O 在 45℃过量的吡啶中反应 24h，得到分子组成为 (C$_5$H$_5$NH)[(C$_5$H$_5$N)$_2$Mo$_8$O$_{26}$] 的白色粉末，该粉末遇水失去阴离子上的吡啶分子，生成组成为 [(C$_5$H$_5$N)$_4$Mo$_8$O$_{26}$] 的化合物。将 (C$_5$H$_5$NH)[(C$_5$H$_5$N)$_2$Mo$_8$O$_{26}$] 溶于体积比为 2∶1 的二甲亚砜和吡啶的混合溶液中，可得到适于作结构分析的单晶，其分子组成为 (C$_5$H$_5$NH)$_4$

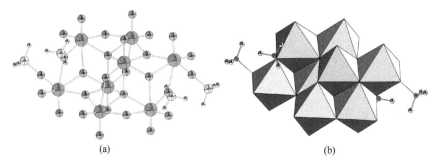

图 8-37　化合物 $Na_4[Mo_8O_{24}(OCH_3)_4] \cdot 8CH_3OH$ 中阴离子结构

$[(C_5H_5N)_2Mo_8O_{26}] \cdot 2Me_2SO^{[74]}$，化合物阴离子结构如图 8-38 所示。吡啶分子中的氮原子直接与阴离子中的钼原子配位，这是首个由非氧原子直接与钼原子配位的多钼氧簇阴离子。

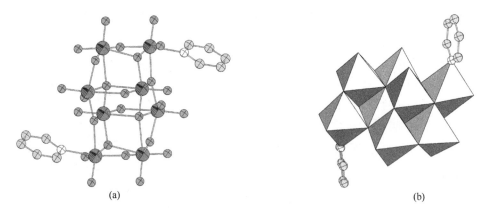

图 8-38　化合物 $(C_5H_5NH)_4[(C_5H_5N)_2Mo_8O_{26}] \cdot 2Me_2SO$ 阴离子结构

向 $[MoO_2(Sal)_2]$ 的甲酸溶液中加入正丙胺，加热回流 30min 或者向 $(NH_4)_6[Mo_7O_{24}]$ 的水溶液中加入水杨醛缩丙亚胺的甲醇溶液，用浓盐酸酸化，加热回流 30min，可以得到分子组成为 $[NH_3Pr]_2[Mo_8O_{22}(OH)_4(OC_6H_4CH=NPr-2)_2] \cdot 6MeOH$ 的化合物[75]。

$[MoO_2(Sal)_2] + NH_2Pr[NH_4]_6Mo_7O_{24} \cdot 4H_2O + HOC_6H_4CH=NPr-2 \longrightarrow [NH_3Pr]_2[Mo_8O_{22}(OH)_4(OC_6H_4CH=NPr-2) \cdot Mo_8O_{22}(OH)_4(OC_6H_4CH=NPr-2)] \cdot 6MeOH$

在该化合物的分子结构中，水杨醛缩丙亚胺中的氧原子与 $[\gamma-Mo_8O_{22}(OH)_4]^{2-}$ 中的五配位钼原子配位（图 8-39）。

MoO_2Cl_2 和蛋氨酸 $[CH_3S(CH_2)_2CHNH_2CO_2H]$ 的甲醇溶液用吗啉甲醇溶液中和到 pH=6，滤出白色沉淀，用水重结晶后可得到分子组成为 $[Hmorph]_4[Mo_8O_{24}(OH)_2(MetO)_2] \cdot 4H_2O^{[75]}$ 的化合物。图 8-40 为化合物阴离子的结构，在阴离子中，有机配体通过氧原子与 $[\gamma-Mo_8O_{24}(OH)_2]^{2-}$ 中的五配位钼原子配位。

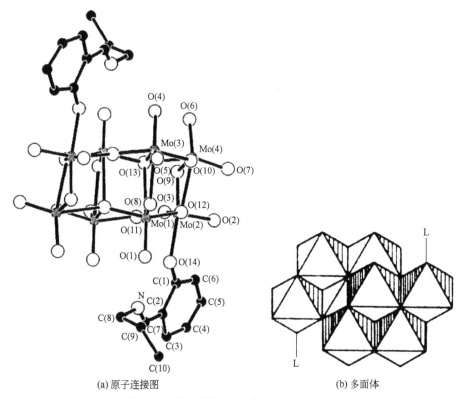

(a) 原子连接图 (b) 多面体

图 8-39 化合物 $[NH_3Pr]_2Mo_8O_{22}(OH)_4(OC_6H_4CH=NPr-2) \cdot 6MeOH$ 中阴离子结构

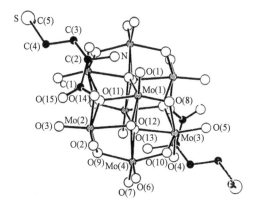

图 8-40 化合物 $[Hmorph]_4[Mo_8O_{24}(OH)_2(MetO)_2] \cdot 4H_2O$ 阴离子结构

三氧化钼 MoO_3 与咪唑、1-甲基咪唑或 2-甲基咪唑在水溶液中回流 12h，可以得到分子组成分别为 $[H_2im]_2[Mo_8O_{26}(Him)_2]$、$[1-Hmim]_4[Mo_8O_{26}(1-mim)_2] \cdot 2H_2O$ 和 $[2-Hmim]_4[Mo_8O_{26}(2-mim)_2] \cdot H_2O$ 的化合物。将 MoO_3 和咪唑在 DMF 和水的混合溶液中回流 12h，得到化合物的分子组成为

$[Me_2NH_2]_4[Mo_8O_{26}(Him)_2] \cdot 3H_2O^{[76]}$。在这些化合物中，$Mo_8O_{26}^{4-}$ 均具有 γ 构型，咪唑（或衍生物）分子中的氮原子与 $[\gamma-Mo_8O_{24}]^{4-}$ 中五配位 MoO_5 四方锥构型的底部配位。在手性配位赖氨酸（D 和 L 构型）的存在下酸化 $Na_2MoO_4 \cdot 2H_2O$ 水溶液，可分别得到 $Na_2[Mo_8O_{26}(D-LySH_2)_2] \cdot 8H_2O$ 和 $Na_2[Mo_8O_{26}(L-LySH_2)_2] \cdot 8H_2O^{[77]}$。两个化合物中分别结合左旋配体和右旋配体的两个阴离子结构很像（图 8-41），在每一个阴离子中，八钼氧簇阴离子骨架具有 γ 构型，每个阴离子中的两个质子化的赖氨酸分子分别与阴离子中两个五配位的钼原子结合。

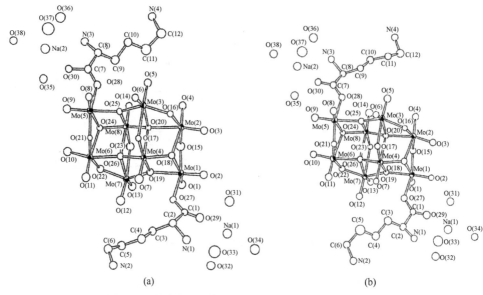

图 8-41 化合物 $Na_2[Mo_8O_{26}(D-LySH_2)_2] \cdot 8H_2O$ 和
$Na_2[Mo_8O_{26}(L-LySH_2)_2] \cdot 8H_2O$ 中阴离子结构

用 MoO_3、吡唑和硒酸在水中回流，通过溶剂扩散生长晶体可得到化合物 $[(Hpyr)_4][(Pyr)_2Mo_8O_{26}] \cdot CH_3COCH_3 \cdot 2H_2O^{[78]}$，化合物阴离子骨架具有 γ 结构，两个吡唑环上的各一个氮原子，分别与 $[\gamma-Mo_8O_{24}]^{4-}$ 中两个五配位的钼原子配位。由 MoO_3 和 $4,4'$-二吡啶胺在水热 170℃下反应 41h，得到组成为 $[4,4'\text{-dpaH}]_2[Mo_8O_{26}(4,4'\text{-dpa})_2]$ 的化合物[79]。同样，在该化合物中，两个 $4,4'$-二吡啶胺分子中分别用其中一个吡啶环上的氮原子与 $[\gamma-Mo_8O_{24}]^{4-}$ 配位。用 $N_2H_4 \cdot H_2SO_4$、$[NH_4]_6Mo_7O_{24} \cdot 4H_2O$ 和 CH_3COONH_4 在水溶液中温和反应，得到化合物 $[NH_4]_6[(CH_3CO)(Mo_8O_{28})] \cdot 4H_2O^{[80]}$。该化合物阴离子与早先报道的 $[(HCO)_2(Mo_8O_{28})]^{6-}$ 结构类似，只是用乙酸根取代了甲酸根和 $[\gamma-Mo_8O_{24}]^{4-}$ 配位。钼酸 H_2MoO_4 和咪唑在 60℃水中搅拌澄清后，加入

$Cu(CH_3COO)_2 \cdot H_2O$,可得到分子组成为$[Cu(imi)_2(H_2O)_4][Himi]_2$
$[(imi)_2Mo_8O_{26}]$的化合物,在该化合物阴离子中,两个咪唑分子分别用一个氮
原子和$[\gamma-Mo_8O_{24}]^{4-}$上的五配位钼原子配位[81],阴离子结构如图8-42所示。

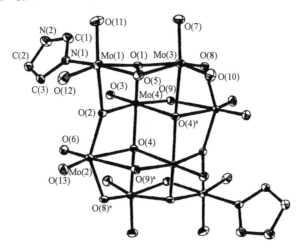

图8-42 化合物$[Cu(imi)_2(H_2O)_4][Himi]_2[(imi)_2Mo_8O_{26}]$
中阴离子结构

H_2MoO_4、$Cu(OOCCH_3)_2 \cdot H_2O$、咪唑在水热条件下160℃反应6d,得到
结构单元组成为$[Cu(imi)_2]_4[(imi)_2Mo_8O_{26}] \cdot 4H_2O$的化合物[82]。尽管该化
合物和前述化合物所用原料一样[81],但由于反应条件的差异,导致化合物结构
差别很大。在化合物$[Cu(imi)_2]_4[(imi)_2Mo_8O_{26}] \cdot 4H_2O$中,阴离子除了两个
五配位的钼原子分别接受来自两个咪唑环上的一个氮原子配位外,还通过四个端
基氧原子分别与四个$Cu(imi)_2^+$配离子配位,其中两个与MoO_5N八面体上的端
基氧原子配位的$Cu(imi)_2^+$,又分别与相邻的两个$[\gamma-Mo_8O_{26}(imi)_2]^{4-}$配位,形
成一维链状结构(图8-43)。

用钼酸H_2MoO_4和咪唑在80℃下水溶液中反应0.5h,得到分子组成为
$(Himi)_4[Mo_8O_{26}(imi)_2] \cdot 4H_2O$的化合物[83],在该化合物中,由于没有金属
离子的参与,阳离子为质子化的咪唑,阴、阳离子之间通过静电引力和氢键作用
结合,在阴离子中,两个咪唑分子分别用一个氮原子与五配位的钼原子配位。烟
酸、钼酸和咪唑在100℃水中搅拌回流,用醋酸调pH值到4.3,可得到组成为
$[C_5H_5N_2]_4[(C_6H_5NO)_2(Mo_8O_{26})]$的化合物,在该化合物阴离子中,两个烟
酸根分别用羧基上的一个氧原子与$[\gamma-Mo_8O_{26}]^{4-}$上的五配位钼原子相连[84,85],
阴离子结构如图8-44所示。

钼酸钠$Na_2MoO_4 \cdot 2H_2O$水溶液用盐酸酸化,加入甘氨酸二肽(Gly-Gly),
搅拌后放置,可得到无色晶体,其单元组成为$\{[Na_2(H_2O)_4]_2[\gamma-Mo_8O_{26}$

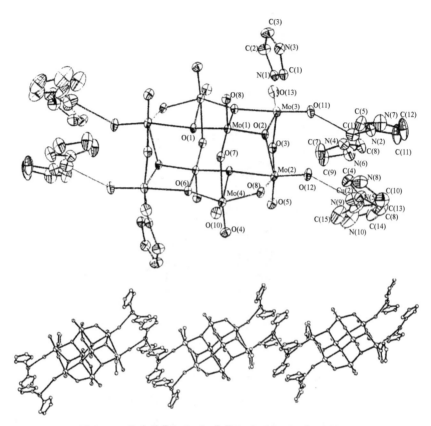

图 8-43 化合物$[Cu(imi)_2]_4[(imi)_2Mo_8O_{26}]\cdot 4H_2O$
阴离子结构单元和一维链状结构

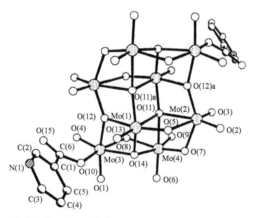

图 8-44 化合物$[C_3H_5N_2]_4[(C_6H_5NO)_2(Mo_8O_{26})]$阴离子结构

$(Gly-Gly)_2]\}$。在该化合物中,两个甘氨酸二肽 Gly-Gly 分子分别用羧基上的一个氧原子与$[\gamma-Mo_8O_{26}]^{4-}$中的五配位钼原子配位,形成$\gamma-[Mo_8O_{26}(Gly-$

Gly)$_2$]$^{4-}$,然后,Gly-Gly 中的非羧基氧原子与桥联基团 [Na$_2$(H$_2$O)$_4$]$^{2+}$ 中的 Na$^+$ 配位,形成二维无限平面结构[86]。K$_2$MoO$_4$、KCl、H$_2$NCH$_2$COOH 和 N$_2$H$_4$·2HCl 在 413K 反应 4d,得到组成为 K$_4$[Mo$_8$O$_{26}$(NH$_3$CH$_2$COO)$_2$]·6H$_2$O 的化合物[87]。在该化合物阴离子中,两个氨基乙酸分子分别用羧基上的一个氧原子与 [γ-Mo$_8$O$_{26}$]$^{4-}$ 中五配位的钼原子配位,生成 γ-[Mo$_8$O$_{26}$(NH$_3$COO)$_2$]$^{4-}$。由 MoO$_2$(acac)$_2$、2-氨基-6-甲基吡啶(amp),在水和 DMF 的混合溶剂中 60℃下反应 4h,得到分子组成为 [H-amp]$_4$[Mo$_8$O$_{26}$(DMF)$_2$]·2H$_2$O 的化合物[88]。在化合物中,质子化的 2-氨基-6-甲基吡啶通过静电引力和氢键与 γ-[Mo$_8$O$_{26}$(DMF)$_2$]$^{4-}$ 相连,在 [γ-Mo$_8$O$_{26}$(DMF)$_2$]$^{4-}$ 中,两个 DMF 分子通过分子内的氧原子与 [γ-Mo$_8$O$_{26}$]$^{4-}$ 中的五配位钼原子配位,形成图 8-45 所示结构的阴离子。

图 8-45 [γ-Mo$_8$O$_{26}$(DMF)$_2$]$^{4-}$ 结构

由钼酸钠、1,2-二胺基苯和草酸,在水热 140℃下反应 84h,得到分子组成为 [Hbenzimi]$_4$[(benzimi)$_2$Mo$_8$O$_{26}$]·2H$_2$O 的化合物[89],该化合物中的苯并咪唑来源于如下反应:

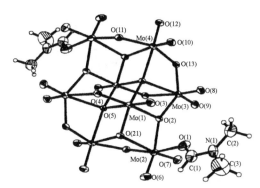

化合物中质子化的苯并咪唑通过静电引力和氢键与 [(benzimi)$_2$Mo$_8$O$_{26}$] 相作用。在 [(benzimi)$_2$Mo$_8$O$_{26}$]$^{4-}$ 中,两个苯并咪唑分子分别用咪唑环上的一个氮原子和 [γ-Mo$_8$O$_{26}$]$^{4-}$ 中五配位的钼原子配位。

8.1.4 δ 异构体

八钼氧簇 δ 异构体,可以通过水热法和普通溶液法来合成,但到目前为止,报道的 [δ-Mo$_8$O$_{26}$]$^{4-}$ 构成的化合物还很少。

(1) 合成

[(RhCp*)$_4$Mo$_4$O$_{16}$]·2H$_2$O 和 CH$_3$SH30%的甲醇溶液,在氩气保护下在甲醇中搅拌回流 3.5h,经过滤和乙腈重结晶,得到分子组成为[(RhCp*)$_2$(μ_2-SCH$_3$)$_3$]$_4$[Mo$_8$O$_{26}$]·2CH$_3$CN 的化合物,在该化合物中,八钼氧簇阴离子 [Mo$_8$O$_{26}$]$^{4-}$ 具有 δ 构型[90]。

(2) 结构

δ 构型八钼氧簇阴离子 [δ-Mo$_8$O$_{26}$]$^{4-}$ 的原子连接图和多面体图如图 8-46 所示。

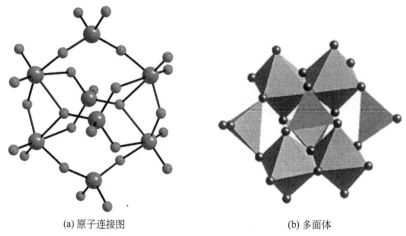

(a) 原子连接图　　(b) 多面体

图 8-46　[δ-Mo$_8$O$_{26}$]$^{4-}$ 原子连接图和多面体

由结构图可知,阴离子具有 C$_{2h}$ 对称性。阴离子中的六个钼原子通过 Mo—O 键相连,构成一个 Mo$_6$O$_6$ 环,在 Mo$_6$O$_6$ 环上有两个钼原子与两个端基氧原子相连,构成四配位四面体构型,这两个钼原子在环上处于对位;另外四个钼原子,除分别与两个端基氧原子相连以外,两两之间还通过一个桥氧原子相连,并通过该桥氧原子与环两侧的一个钼氧四面体相连,同时,每个钼原子还分别通过一个桥氧原子与位于环另一侧的钼原子相连;位于环两侧的两个钼原子除连接一个端基氧原子外,还分别通过一个三重桥氧原子和两个二重桥氧与四个六配位的钼原子相连,构成四配位四面体构型。阴离子中的氧原子根据其配位环境的不同,可分为三类,分别是与一个钼原子相连的端基氧原子(O$_t$),共有 14 个;与两个钼原子相连的二重桥氧原子(O$_b$),共有 10 个;与三个钼原子相连的三重桥氧原子(O$_c$),共有 2 个。阴离子中的钼原子可分为三类:位于环上的六配位钼原子 Mo$_I$,分别连接两个端基氧原子、三个二重桥氧原子和一个三重桥氧原子;位于 Mo$_6$O$_6$ 环上的四配位钼原子 Mo$_{II}$,分别连接两个端基氧原子和两个二重桥氧原子;位于环两侧的两个四配位钼原子,分别连接一个端基氧原子、两个二重桥氧原子和一个三重桥氧原子。

(3) 相关化合物

由 $CuSO_4 \cdot 5H_2O$、MoO_3 和 4,4'-联吡啶,在水热 200℃下反应 96h,可制得分子组成为 $[\{Cu(4,4'-bpy)\}_4Mo_8O_{26}]$ 的化合物,在该化合物中,$Mo_8O_{26}^{4-}$ 具有 δ 构型[91]。三氧化钼和 2,4,6-三-4-吡啶基三嗪(tptz)在水溶液中调 pH=5.5,在水热 180℃下反应 48h,得到分子组成为 $(H_2tptz)_2[\delta-Mo_8O_{26}] \cdot 2H_2O$ 的化合物[92],化合物中阴、阳离子之间通过静电引力和氢键作用结合,阴离子具有 δ 构型。$NiSO_4 \cdot 6H_2O$、$(NH_4)_6Mo_7O_{24} \cdot 4H_2O$、2,2'-联吡啶和水用盐酸调 pH 值到 5.5,在 160℃水热条件下反应 194h,得到分子组成为 $[Ni(2,2'-bpy)_3]_2[\delta-Mo_8O_{26}]$ 的化合物[93],该化合物中,阴离子具有 δ 构型。化合物分子内 $[\delta-Mo_8O_{26}]^{4-}$ 和 $[Ni(2,2'-bpy)_3]^{2+}$ 之间通过静电引力和氢键结合,结构如图 8-47 所示。

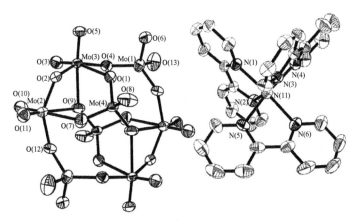

图 8-47 化合物 $[Ni(2,2'-bpy)_3]_2[\delta-Mo_8O_{26}]$ 结构

$(NH_4)_6Mo_7O_{24} \cdot 4H_2O$、$Cu(CH_3COO)_2 \cdot H_2O$、四-2-吡啶基吡嗪(tpyprz)、4,4'-联苯二磷酸和水在水热 200℃条件下反应 48h,得到结构单元组成为 $[\{Cu_2(tpyprz)\}_2Mo_8O_{26}] \cdot 2H_2O^{[6]}$ 的化合物,在该化合物中,$Mo_8O_{26}^{4-}$ 具有 δ 构型,阴离子通过 Mo_6O_6 环上的两个 MoO_4 四面体和两个 MoO_6 六面体上的端基氧分别与两个配阳离子配位(图 8-48),进而通过配阳离子构成二维层状结构。

8.1.5 ε 异构体

八钼氧簇 ε 异构体报道很少,目前仅见于由水热合成制得的 $[\epsilon-Mo_8O_{26}]^{4-}$ 构成化合物的报道[91]。

由 $NiCl_2 \cdot 6H_2O$、MoO_3 和 4,4'-联吡啶在水热 200℃的条件下反应 96h,可得到组成为 $[\{Ni(H_2O)_2(4,4'-bpy)_2\}_2Mo_8O_{26}]$ 的结构单元。在该化合物中,阴

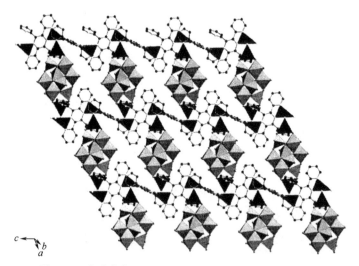

图 8-48　化合物 $[\{Cu_2(tpyprz)\}_2Mo_8O_{26}]\cdot 2H_2O$ 结构

离子具有 ε 构型。

八钼氧簇 ε 异构体 $[\varepsilon\text{-}Mo_8O_{26}]^{4-}$ 的原子连接图和多面体如图 8-49 所示。由结构图可知，阴离子具有 C_{2h} 对称性。阴离子六个钼原子通过桥氧原子相连构成 Mo_6O_6 环，环上的每一个钼原子连接两个端基氧原子，另有四个位于环两侧的桥氧原子分别与环上两个相邻的钼原子相连，其中位于对位的两个钼原子分别与两个侧面的桥氧原子相连，就构成了由四个 MoO_5 四方锥和两个 MoO_6 八面体通过共边和共顶点构成的环。另外两个钼原子分别位于环的两侧，除连接两个端基氧原子外，还分别连接两个位于环侧面的桥氧原子和一个位于环上的桥氧原子，构成五配位四方锥构型。按照配位环境分，阴离子中的氧原子可分为三类：①只与一个钼原子相连的端基氧原子（O_t），共有 16 个；②连接两个钼原子的二重桥氧原子（O_b），共有 4 个；③连接三个钼原子的三重桥氧原子（O_c），共有 6 个。根据连接氧原子类型的不同，钼原子可分为三类：①八面体构型钼原

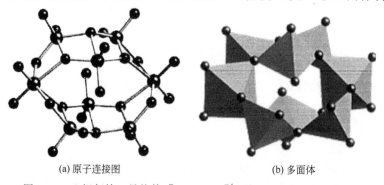

(a) 原子连接图　　　　　　(b) 多面体

图 8-49　八钼氧簇 ε 异构体 $[\varepsilon\text{-}Mo_8O_{26}]^{4-}$ 的原子连接图和多面体

子，连接两个端基氧原子、两个二重桥氧原子和两个三重桥氧原子；②环上的四方锥钼原子，连接两个端基氧原子、一个二重桥氧原子和两个三重桥氧原子；③侧面的四方锥钼原子，连接两个端基氧原子和三个三重桥氧原子。

8.1.6 ζ异构体

目前所报道的八钼氧簇ζ异构体[ζ-Mo_8O_{26}]$^{4-}$构成的化合物由水热方法制备。首例含有[ζ-Mo_8O_{26}]$^{4-}$的化合物在2004年报道[94]。由$CuSO_4·5H_2O$、MoO_3、四-4-吡啶吡嗪，在水热120℃条件下反应48h，得到结构单元为[{Cu_2(tpyrpyz)}$_2$ {ζ-Mo_8O_{26}}]·$7H_2O$的化合物，在该化合物中，八钼氧簇阴离子具有ζ构型[94]，结构如图8-50所示。

阴离子中的6个钼原子通过桥氧原子相连构成Mo_6O_6环，环上的每一个钼原子除与两个桥氧原子相连外，均连接两个端基氧原子，另外，六个桥氧原子分布于环的两侧，每侧有两个三重桥氧原子和一个二重桥氧原子，四个三重桥氧原子分别连接两个位于环上的钼原子和一个位于侧面的钼原子；两个二重桥氧原子分别连接一个环上钼原子和1个侧面的钼原子，构成环上四个六配位八面体和两个五配位四方锥和两侧两个五配位四方锥。根据[ζ-Mo_8O_{26}]$^{4-}$内氧原子的配位环境，氧原子可分为三类：①连接一个钼原子的端基氧原子（O_t），共14个；②连接两个钼原子的二重桥氧原子（O_b），共6个；③连接三个钼原子的三重桥氧原子（O_c），共6个。阴离子中的钼原子，按照键合氧原子种类的不同，也可分为三类：连接两个端基氧原子、两个二重桥氧原子和两个三重桥氧原子的六配位八面体构型钼原子，位于Mo_6O_6环上，共4个；连接两个端氧原子、一个二重桥氧原子和两个三重桥氧原子的五配位四方锥构型钼原子，位于Mo_6O_6环上，共2个；连接一个端基氧原子、一个二重桥氧原子和三个三重桥氧原子的五配位四方锥构型钼原子，位于Mo_6O_6环的两侧，共2个。整个[ζ-Mo_8O_{26}]$^{4-}$由四个MoO_6八面体和四个MoO_5四方锥构成。

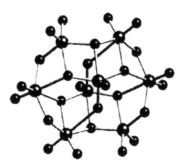

图8-50 八钼氧簇ζ异构体[ζ-Mo_8O_{26}]$^{4-}$原子连接图

由钼酸钠$NaMoO_4·2H_2O$、醋酸银$Ag(CH_3COO)_2$、四-2-吡啶吡嗪和HF在水热170℃下反应48h，得到结构单元组成为[{Ag_4(tpyprz)$_2$(H_2O)}(Mo_8O_{26})]的化合物[95]，在该化合物中，阴离子具有ζ构型。阴离子通过两两相邻的四方锥和八面体上的一个端基氧原子与银离子配位。

8.1.7 η 构型

Zubieta 报道了含有八钼氧簇 η 异构体的化合物 $[\{Cu(t\text{-}Bu_2bpy)\}_2Mo_8O_{26}]$ 的合成[94]，$[\eta\text{-}Mo_8O_{26}]^{4-}$ 结构如图 8-51 所示。

图 8-51　八钼氧簇 η 异构体 $[\eta\text{-}Mo_8O_{26}]^{4-}$ 结构

阴离子由六个八面体和两个四方锥构成。六个钼原子通过桥氧相连构成一个 Mo_6O_6 环，环上的每一个钼原子除与两个相邻钼原子相连的桥氧原子连接外，还分别连接两个端基氧原子，另有六个桥氧原子分别于环的两侧，每个氧原子分别与环上的两个钼原子相连，另外两个钼原子位于环的两侧，除分别与三个位于两侧的氧原子相连外，还与 Mo_6O_6 环上的一个氧原子相连，同时连接一个端基氧原子，整个阴离子由六个共边相连的钼氧八面体 MoO_6 形成一个八面体环，两个 MoO_5 四方锥位于环的两侧通过桥氧原子与八面体环相连。根据连接原子个数的不同，阴离子中的氧原子可分为三类：①只连接一个钼原子的端基氧原子（O_t），共 14 个；②连接两个钼原子的桥氧原子（O_b），共 4 个；③连接三个钼原子的桥氧原子（O_c），共 8 个。钼原子可分为位于环上的八面体钼原子和位于环侧的四方锥钼原子。

8.1.8 θ 构型

2004 年首次报道了含有八钼氧簇 θ 结构阴离子的化合物 $[\{Ni(phen)_2\}_2(Mo_8O_{26})]$[96]。但由于当时报道 η 结构的八钼氧簇异构体的文献尚在出版之中而被定为 η 结构。随后，又有类似的含有 $[\theta\text{-}Mo_8O_{26}]^{4-}$ 的化合物报道[97]。$[\theta\text{-}Mo_8O_{26}]^{4-}$ 结构如图 8-52 所示。

阴离子由四个 MoO_6 八面体、两个 MoO_5 四方锥和两个 MoO_4 四面体，通过共边和共顶点相连而成。$[\theta\text{-}Mo_8O_{26}]^{4-}$ 构型和 α 异构体构型很像，只是 α 异构体与侧面 MoO_4 四面体相连的三个三重桥氧原子中的一个只与一个环上的钼原子配位，在 α 构型中位于环上的六个八面体就有位于对位的两个变成了四方锥。因此 $[\theta\text{-}Mo_8O_{26}]^{4-}$ 就是由两个钼氧八面体和一个钼氧 MoO_5 四方锥通过共边相连，两组 Mo_3O_{12} 基团通过共顶点相连形成一个六

图 8-52　八钼氧簇 θ 异构体 $[\theta\text{-}Mo_8O_{26}]^{4-}$ 结构

元环，两个 MoO_4 四面体通过共顶点分别位于环的两侧与环相连。到目前为止，$[\theta\text{-}Mo_8O_{26}]^{4-}$ 是包含多面体类型最多的一种异构体。阴离子中氧原子的类型可分为三类：①端基氧原子（O_t），共 14 个；②二重桥氧原子（O_b），共 8 个；③三重桥氧原子（O_c），共 4 个。阴离子中的钼原子可分为三类：四面体构型钼原子，共 2 个；四方锥构型钼原子，共 2 个；八面体构型钼原子，共 4 个。

8.1.9 八钼氧簇异构体比较

上面分别介绍了八钼氧簇阴离子 α、β、γ、δ、ε、ζ、η、θ 共八种异构体的结构特征，为了更好地了解这些阴离子的结构，表 8-3 列出了各种异构体的结构特征。

表 8-3 $[Mo_8O_{26}]^{4-}$ 异构体结构特征

异构体	多面体单元	桥氧类型个数	异构体	多面体单元	桥氧类型个数
α	$6O_h, 4T_d$	$14O_t, 6O_b, 6O_c$	ε	$2O_h, 6sp$	$16O_t, 4O_b, 6O_c$
β	$8O_h$	$14O_t, 6O_b, 4O_c, 2O_d$	ζ	$4O_h, 4sp$	$14O_t, 6O_b, 6O_c$
γ	$6O_h, 2sp$	$14O_t, 6O_b, 4O_c, 2O_d$	η	$6O_h, 2sp$	$14O_t, 4O_b, 8O_c$
δ	$4O_h, 4T_d$	$14O_t, 10O_b, 2O_c$	θ	$4O_h, 2sp, 2T_d$	$14O_t, 8O_b, 4O_c$

注：O_h 为八面体，sp 为四方锥，T_d 为四面体。

8.2 八金属氧簇 M_8O_{30}

八金属氧簇的另一种构成类型 M_8O_{30} 是最近才被发现的一种新的金属氧簇阴离子结构类型，尽管还没有独立结构阴离子存在的报道，但该金属氧簇阴离子已在担载结构中多次出现，目前，有钼、钨两种金属可以形成该种结构类型的八金属氧簇[98,99]。

8.2.1 八钼氧簇

(1) 合成

$(NH_4)_6[Mo_7O_{24}] \cdot 4H_2O$ 和 $[Mn(CO)_5]Br$ 在一定条件下反应后过滤，滤液室温避光缓慢挥发得到红色针状晶体[98]，以 $[Mn(CO)_5]Br$ 为基准计算产率为 28%，化合物分子组成为 $[(NH_4)_4]\{[Mo_8O_{30}H_6][Mn(CO)_3]_2\} \cdot 12H_2O$。将 $[Mn(CO)_5]Br$ 换成 $[Re(CO)_5]Cl$，室温避光缓慢挥发可得到分子组成为 $[(NH_4)_4]\{[Mo_8O_{30}H_6][Re(CO)_3]_2\} \cdot 14H_2O$ 的红色针状晶体。

(2) 结构

化合物 $[(NH_4)_4]\{[Mo_8O_{30}H_6][Mn(CO)_3]_2\} \cdot 12H_2O$ 由 4 个 NH_4^+、1 个

图 8-53 {[$H_6Mo_8O_{30}$][$Re(CO)_3$]$_2$}$^{4-}$ 结构

{[$H_6Mo_8O_{30}$][$Mn(CO)_3$]$_2$}$^{4-}$ 和 12 个结晶水组成。八钼氧簇 [Mo_8O_{30}] 结构如图 8-53 所示，整个八钼氧簇构筑块可分为三层，上下各 1 组 Mo_3O_{13} 三金属簇通过中间 2 个 MoO_6 八面体相连，阴离子存在 1 个晶体学反演中心。

{[$H_6Mo_8O_{30}$][$Mn(CO)_3$]$_2$}$^{4-}$ 与 {[$H_6Mo_8O_{30}$][$Re(CO)_3$]$_2$}$^{4-}$ 同构，2 个羰基锰基团分别通过与上、下 2 个三金属簇中的 3 个二重桥配位与八钼氧簇 [$H_6Mo_8O_{30}$] 阴离子相连；也可以看作是 2 个由三羰基锰基团和 3 个 Mo 原子构建的 {Mo_3MO_4} 立方烷单元通过 2 个 MoO_6 八面体连接起来。氧原子按其配位环境的不同分为 4 类：14 个只与 1 个 Mo 原子配位的端氧 O_t，8 个与 2 个 Mo 原子配位的二重桥氧 μ_2-O_b，2 个与 3 个 Mo 原子配位的三重桥氧 μ_3-O_{b1}，6 个与 2 个 Mo 原子和 1 个 Mn 原子配位的三重桥氧 μ_3-O_{b2}。

8.2.2 八钨氧簇

(1) 合成

$Mn(CO)_5Br$、$Na_2WO_4 \cdot 2H_2O$ 和冰醋酸按一定的比例混合于溶剂中，一定条件下反应，采用溶液挥发法培养单晶，三周后得到橘黄色块状晶体。以 $Mn(CO)_5Br$ 计产率为 35%，化合物分子组成为 [$Na(H_2O)_5$]$_2H_6${[$H_2W_8O_{30}$][$Mn(CO)_3$]$_2$} $\cdot 13H_2O^{[99]}$。用 $Re(CO)_5Br$ 代替 $Mn(CO)_5Br$，可得到分子组成为 H_2[$Na(H_2O)_5$]$_2$[$Na(H_2O)_4$]$_2$[$Na(H_2O)_2$]$_2${[$H_2W_8O_{30}$][$Re(CO)_3$]$_2$} 的黄绿色块状晶体，以 $Re(CO)_5Br$ 计产率为 20%。$Mn(CO)_5Br$、$Na_2WO_4 \cdot 2H_2O$、$Mn(Ac)_2 \cdot 4H_2O$ 和冰醋酸按一定比例混合于混合溶剂中，一定条件下反应，三周后得到黄色针状晶体。以 $Mn(CO)_5Br$ 计产率为 38%，其分子组成为 [$Na(H_2O)_5$]$_2$[$Na_2(\mu_2$-$H_2O)_2(H_2O)_4$]$_2$[$Mn(H_2O)_2$]$_2${[$H_2W_8O_{30}$][$Mn(CO)_3$]$_2$}。用 $AgNO_3$ 代替 $Mn(Ac)_2 \cdot 4H_2O$，可得到分子组成为 [$Mn(H_2O)_2$]$_2H_6${[$H_2W_8O_{30}$][$Mn(CO)_3$]$_2$} $\cdot 24H_2O$ 的黄色针状晶体。以 $Mn(CO)_5Br$ 计产率为 35%。

(2) 结构

X 射线单晶衍射分析结果表明，化合物均含有八钨氧簇阴离子结构单元 [W_8O_{30}]，该结构单元与 [Mo_8O_{30}] 同构。在八钨氧簇与羰基金属构成的阴离子 {[$H_2W_8O_{30}$][$Mn(CO)_3$]$_2$}$^{8-}$ 中，羰基金属与八钨氧簇骨架的结合方式与同八钼氧簇的结合方式相同。

参考文献

[1] Fuchs J, Hartl H. Anion structure of tetrabutylammonium octamolybdate $[N(C_4H_9)_4]_4Mo_8O_{26}$. Angew Chem Int Ed Engl, 1976, 15 (6): 375-376.

[2] Klemperer W G, Shum W. Synthesis and interconversion of the isomeric α-and β-$[Mo_8O_{26}]^{4-}$ ions. J Am Chem Soc, 1976, 98 (25): 8291-8293.

[3] Day V W, Fredrich M F, Klemperer W G, Shum W. Structural and dynamic stereochemistry of α-$[Mo_8O_{26}]^{4-}$. J Am Chem Soc, 1997, 99 (3): 952-953.

[4] Lu S, Huang J, Huang Z, Huang J. Crystal structure of $H_2(Et_4N)_2[Mo_8O_{26}]$. J Struct Chem, 1988, 7 (3): 23-26.

[5] Wu C D, Zhan X P, Lu C Z, Zhuang H H. Hydrothermal assembly and structural characterization of an octamolybdate supported copper (I) tetramer: $[Cu_2(1,10\text{-phen})_2(4,4'\text{-bpy})]_2[Mo_8O_{26}] \cdot 4H_2O$. Chin J Struct Chem, 2002, 21 (5): 525-529.

[6] Burkholder E, Zubieta J. Two-dimensional oxides constructed from octamolybdate clusters and M^{n+}/tetrapyridylpyrazine subunits(M=Co, Ni, $n=2$; M=Cu, $n=1$). Inorga Chim Acta, 2005, 358: 116-122.

[7] Hagrman D Hagrman P, Zubieta J. Polyoxomolybdate clusters and copper-organonitrogen complexes as building blocks for the construction of composite solids. Inorga Chim Acta, 2000, 300-302: 212-224.

[8] Hagrman D, Sangregorio C, O'Connor C J, Zubieta J. Solid state coordination chemistry: two-dimensional oxides constructed from polyoxomolybdate clusters and copper-organoamine subunits. J Chem Soc Dalton Trans, 1998: 3707-3709.

[9] Yang W B, Lu C Z, Zhuang H H. Hydrothermal synthesis and structures of three new copper complexes: $[\{Cu(2,2'\text{-bipy})\}_2(\beta\text{-}Mo_8O_{26})]$, $[\{Cu(py)_3\}_2\{Cu(py)_2\}_2(\alpha\text{-}Mo_8O_{26})]$ and $[Cu(py)_2]_4[(SO_4)Mo_{12}O_{36}]$. J Chem Soc Dalton Trans, 2002: 2879-2884.

[10] Rarig R S Jr, Zubieta J. Octamolybdate subunits as building blocks in the hydrothermal synthesis of organically templated mixed metal oxides: the synthesis and X-ray characterization of $[Cu_2Mo_4O_{13}(3,3'\text{-bipy})_2] \cdot H_2O$, $[CuMo_4O_{13}(\text{hdipyreth})]$ and $[Cu(dpp)]_2[Cu_2(\alpha\text{-}Mo_8O_{26})(dpp)_2] \cdot 2H_2O$($3,3'$-bipy=$3,3'$-bipyridine; dipyreth=1,2-bis(2-pyridyl)ethylene; dpp=$4,4'$-trimethylenedipyridine). Polyhedron, 2003, 22: 177-188.

[11] Shi Y P, Yang W, Xue G L, Hu H M, Wang J W. A novel crystal coexisting with two kinds of polyoxomolybdates: $[n\text{-}Bu_4N]_8[Mo_6O_{19}]_2[\alpha\text{-}Mo_8O_{26}]$. J Mol Struct, 2006, 784: 244-248.

[12] Sun C Y, Wang E B, Xiao D R, An H Y, Xu L. The first example of a structure containing both α-and β-octamolybdates: synthesis and structure of a new three-dimensional supramolecular network $[Co(2,2'\text{-bipy})_3]_4[Mo_8O_{26}]_2 \cdot 5H_2O$($2,2'$-bipy=$2,2'$-bipyri-

dine). J Mol Struct, 2005, 741: 149-153.
[13] Ginsberg A P. Inorganic Syntheses. New York: John Wiley&Sons, 1990: 27, 78.
[14] Lindqvist I. Structure of the tetramolybdate ion. Ark Kemi, 1950, 2: 349.
[15] Atovmyan L O, Krasochka O N. X-ray diffraction investigation of the crystals of the octamolybdate $(NH_4)_4Mo_8O_{26} \cdot 4H_2O$. J Struct Chem (USSR), 1972, 13: 319-320.
[16] Weakley T J R. The crystal structure of ammonium β-octamolybdate pentahydrate. Polyhedron, 1982, 1: 17-19.
[17] Vivier H, Barnard J, Djomaa H. Crystal structure of ammonium molybdate tetrahydrate $(NH_4)_4Mo_8O_{26} \cdot 4H_2O$. Rev Chim Miner, 1977, 14: 584-604.
[18] Gatehouse B M. The crystal and molecular structures of $Ce_6Mo_{10}O_{39}$ and $K_2Mo_2O_7 \cdot H_2O$ and the refinement of the "Lindqvist" octamolybdate $(NH_4)_4Mo_8O_{26} \cdot 4H_2O$. J Less-Common Met, 1977, 54: 283-288.
[19] Wilson A J, Mckee V, Penfold B R, Wilkins C J. Structure of tetrakis (dimethylammonium) β-octamolybdate bis (N,N-dimethylformamide), $[NH_2(CH_3)_2]_4[Mo_8O_{26}] \cdot 2C_3H_7NO$, with comments on relationships among octamolybdate anions. Acta Cryst, 1984, C40: 2027-2030.
[20] Yamase T, Ikawa T. Photochemical study of the alkylammonium molybdates. III. Preparation and properties. Bulletin of the Chemical Society of Japan, 1977, 50 (3): 746-749.
[21] Bharadwaj P K, Ohashi Y, Sasada Y. Structure of oxonium tris (triethylammonium) octamolybdate (4−) dihydrate, $(C_6H_{16}N)_3(H_3O)[Mo_8O_{26}] \cdot 2H_2O$. Acta Cryst, 1984, C40: 48-50.
[22] Attanasio D, Bonamico M, Fares V, Suber L. Organic-inorganic charge-transfer salts based on the β-$[Mo_8O_{26}]^{4-}$ isopolyanion: synthesis, properties and X-ray structure. J Chem Soc Dalton Trans, 1992: 2523-2528.
[23] Wang Q Y, Xu X X, Wang X. Structure of bis (dimethylammonium) bis (tetrabutylammonium) β-octamolybdate. Acta Cryst, 1993, C49: 464-467.
[24] Gomez-Garcia C J, Coronado E. An organic-inorganic salt containing mixed-valence TTF chains and the molecular metal oxide cluster $[Mo_8O_{26}]^{4-}$, preliminary spectroscopic, conducting and magnetic properties of the compound $(TTF)_7Mo_8O_{26}$. Synthetic Metals, 1993, 55-57: 1787-1790.
[25] Gutierrez-Zorrilla J M, Yamase T, Sugeta M. Tetrakis (isopropylammonium) β-Octamolybdate (VI). Acta Cryst, 1994, C50: 196-198.
[26] Mccann M, Maddock K, Cardin C, Convery M, Quillet V. Synthesis and reactions of $[Et_3NH]_4[Mo_8O_{26}]$: X-ray crystal structure of $[Et_3NH]_3[NaMo_8O_{26}]$. Polyhedron, 1994, 13 (5): 835-840.
[27] Xu X X, You X Z, Wang X. The crystal structure of tetra (piperidinium) octamolybdate (VI) tetrahydrate. Polyhedron, 1994, 13 (6-7): 1011-1014.
[28] Mccann M, Maddock K, Cardin C, Convery M, Ferguson G. Synthesis and X-ray crys-

tal structure of the triethylammonium magnesium β-octamolybdate (Ⅵ) salt $[Et_3NH]_2$ $[Mg(H_2O)_6Mo_8O_{26}]\cdot 2H_2O$. Polyhedron, 1995, 14 (23-24): 3655-3659.

[29] Kamenar B, Penavic M, Korpar-Colig B, Cindric M. Crystal structure of morpholinium β-octamolybdate tetrahydrate, $(C_4H_{10}NO)_4(Mo_8O_{26})\cdot 4H_2O$. New Crystal Structure, 1995: 535.

[30] Fun H K, Yip B C, Niu J Y, You X Z. Tetrakis (triethylammonium) octamolybdate (Ⅵ) dihydrate. Acta Cryst, 1996, C52: 506-509.

[31] Wang X J, Kang B S, Su C Y, Yu K B, Zhang H X, Chen Z N. Function of the hydrogen bond in the conversion of α-to β-$Mo_8O_{26}^{4-}$ ions formation and structure of tetra (tributylhydrogenammonium) β-octamolybdate. Polyhedron, 1999, 18: 3371-3375.

[32] Gili P, Núñez P, Martin-Zarza P, Lorenzo-Luis P A. A new inorganic-organic hybrid: tetraimidazolium octamolybdate (Ⅵ) containing the β-form of the $[Mo_8O_{26}]^{4-}$ anion. Acta Cryst, 2000, C56: e441-e442.

[33] Wang R Z, Xu J Q, Yang G Y, Li Y F, Xing Y H, Li D M, Liu S-Q, Bu W M, Ye L, Fan Y G. Hydrothermal synthesis and characterization of $[Ni(phen)_3]_2[Mo_8O_{26}]\cdot 2H_2O$. Solid State Sci, 2000, 2: 705-710.

[34] Zheng L M, Wang Y S, Wang X Q, Korp J D, Jacobson A J. Anion-directed crystallization of coordination polymers: syntheses and characterization of $Cu_4(2\text{-pzc})_4(H_2O)_8(Mo_8O_{26})\cdot 2H_2O$ and $Cu_3(2\text{-pzc})_4(H_2O)_2(V_{10}O_{28}H_4)\cdot 6.5H_2O$ (2-pzc=2-pyrazinecarboxylate). Inorg Chem, 2001, 40: 1380-1385.

[35] Wu C D, Lu C Z, Zhuang H H, Huang J S. Bis [4,4′-bipyridinium (2+)] hexacosaoxooctamolybdate. Acta Cryst, 2001, E57: m349-m351.

[36] Lee U, Joo H C, Cho M A. Hexaaquacobalt (Ⅱ) tetrapotassium diethylenediaminetetraacetato (4-) cobaltate (Ⅲ) β-octamolybdate hexahydrate. Acta Cryst, 2002, E58: m599-m601.

[37] Yan Y, Wu C D, Liu J H, Zhang Q Z, Lu C Z. Tetrakis [4-amino-N-(2-hydroxyethyl) pyridinium] hexacosaoxooctamolybdate. Acta Cryst, 2003, E59: m102-m104.

[38] Lapinski A, Starodub V, Golub M, Kravchenko A, Baumer V, Faulques E, Graja A. Characterization and spectral properties of the new organic metal $(BEDT\text{-}TTF)_6(Mo_8O_{26})\cdot 3DMF$. Synthetic Met, 2003, 138: 483-489.

[39] Li J, Qi Y F, Wang E B, Li J, Wang H F, Li Y G, Lu Y, Hao N, Xu L, Hu C W. Synthesis, structural characterization and biological activity of polyoxometallate-containing protonated amantadine as a cation. J Coord Chem, 2004. 57 (9): 715-721.

[40] Xiao D R, Hou Y, Wang E B, Li Y G, Xu L, Hu C W. Hydrothermal synthesis and characterization of a novel polyoxometallate-templated three dimensional supramolecular network. J Coord Chem, 2004, 57 (7): 615-626.

[41] Song L J, Zeng H Y, Dong Z C, Guo G C, Huang J S. Synthesis and structure of paramagnetic β-octamolybdate complex, $[Fe(2,2'\text{-bipy})_3]_2[Mo_8O_{26}]\cdot 6H_2O$. Chin J Struct

Chem, 2004, 23 (2): 135-140.

[42] Chen S M, Lu C Z, Yu Y Q, Zhang Q Z, He X. Bis (tetramethylammonium) hexaaquacobalt (II) β-octamolybdate (VI). Acta Cryst, 2004, E60: m723-m725.

[43] Harrison W T A. Tetrakis (tetramethylammonium) octamolybdate dihydrate. Acta Cryst, 1993, C49: 1900-1902.

[44] Shen Y, Yu L, Niu J Y, Wang J P. Hydrothermal synthesis and crystal structure of a hybrid compound based on β-octamolybdate [Mn(2,2'-bipy)$_3$]$_2$[β-Mo$_8$O$_{26}$] · 4H$_2$O. Chin J Struct Chem, 2005, 24 (10): 1164-1168.

[45] Aguado R, Pedrosa M R, Arnaiz F J. Synthesis and crystal structure of a Mo$_8$O$_{26}^{4-}$ cluster derivative with 4-MePyH$^+$. First β-octamolybdate derivative with π-π stacking. Z Anorg Allg Chem, 2005, 631: 1995-1999.

[46] Gomez-Garcia C J, Coronado E, Triki S, Ouahab L, Delhaes P. First tetrathiafulvalene (TTF) cation-radical salt containing the inorganic polyoxometalate [β-Mo$_8$O$_{26}$]$^{4-}$. Adv Mater, 1993, 5 (4): 283.

[47] Li Q, Zhang S W. A new organic-inorganic charge-transfer salt [C$_{15}$H$_{17}$N$_4$]$_4$[Mo$_8$O$_{26}$]——synthesis, properties and crystal structure. Z Anorg Allg Chem, 2005, 631: 645-648.

[48] Deng Z P, Gao S, Huo L H, Zhao H. Bis (3,5-dimethylpyrazolium) bis (hydroxonium) β-octamolybdate (VI) ethanol disolvate. Acta Cryst, 2005, E61: m2553-m2555.

[49] Feng S S, Lu L P, Zhang H M, Qin S D, Li X M, Zhu M L. A co-crystal of 2-methylbenzimidazole and ammonium octamolybdate. Acta Cryst, 2005, E61: m659-m661.

[50] Muller E A, Sarjeant A N, Norquist A J. (C$_6$H$_{16}$N$_2$)$_2$[Mo$_8$O$_{26}$]: a new β-octamolybdate salt. Acta Cryst, 2005, E61: m730-m732.

[51] Chen L J, Xia C K, Zhang Q Z, Lu C Z. Bis [2-(4-pyridinio)-1H-benzimidazolium] hexacosaoxooctamolybdate (VI). Acta Cryst, 2005, E61: m92-m94.

[52] 鲁晓明, 宗瑞发, 刘顺诚. 苯基氮杂-15-冠-5 对钼同多酸以及钨钼杂多酸阴离子的选择. 高等学校化学学报, 1997, 12: 1911-1916.

[53] 鲁晓明, 刘顺诚, 刘育, 卜显和, 洪少良. N-对 R 苯基氮杂 15 冠 5 八钼多酸钠超分子配合物的合成与结构. 化学学报, 1997, 55: 1009-1018.

[54] Chen Y P, Zhang H H, Ke D M. Structure and two-dimensional correlation infrared spectroscopy study of a new one-dimensional chain compound: (4,4'-Hbpy)$_3$[NaMo$_8$O$_{26}$] (4,4'-bpy)$_2$(H$_2$O)$_4$ (bpy=bipydine). Chin J Struct Chem, 2005, 24 (9): 1033-1038.

[55] Chen Y P, Shen X M, Zhang H H, Huang C C, Cao Y N, Sun R Q. The structure and two-dimensional correlation infrared spectroscopy study of a new 2D polyoxomolybdate complex containing β-octamolybdate linked up by potassium ions: (4,4'-Hbpy)$_2$(K$_2$Mo$_8$O$_{26}$). Vib Spectrosc, 2006, 40: 142-147.

[56] Qin C, Wang X L, Qi Y F, Wang E B, Hu C W, Xu L. A novel two-dimensional β-octamolybdate supported alkaline-earth metal complex: [Ba(DMF)$_2$(H$_2$O)]$_2$[Mo$_8$O$_{26}$] · 2DMF. J Solid State Chem, 2004, 177: 3263-3269.

[57] Wu C D, Lu C Z, Zhuang H H, Huang J S. Hybrid coordination polymer constructed from β-octamolybdates linked by quinoxaline and its oxidized product benzimidazole coordinated to binuclear copper (I) fragments Inorg Chem, 2002, 41: 5636-5637.

[58] Wang R Z, Xu J Q, Yang G Y, Bu W M, Xing Y H, Li D M, Liu S Q, Ye L, Fan Y G. A metal-oxo cluster-supported transition metal complex: synthesis, structure and properties of [Cu(phen)$_2$]$_2$[{Cu(phen)}$_2$Mo$_8$O$_{26}$] · H$_2$O. Polyhedron, 1999, 18: 2971-2975.

[59] Liang Y C, Hong M C, Su W P, Cao R, Chen J T. Inorganic-organic hybrid polymers via hydrothermal syntheses: tetraaquahexakis (pyrazine-2-carboxylato) pertacopper (4+) hexacosaoxooctamolybdate (4−) polymer ({[Cu$_5$(pzca)$_6$(H$_2$O)$_4$][Mo$_8$O$_{26}$]}$_n$; pzca=pyrazine-2-carboxylate) and dicopperdecaoxo (pyrazine) trimolybdenum polymer ([Mo$_3$Cu$_2$O$_{10}$ (pz)]$_n$; pz=pyrazine). Helvetica Chimica Acta, 2001, 84: 3393-3402.

[60] Luo J H, Hong M C, Wang R H, Shi Q, Cao R, Weng J B, Sun R Q, Zhang H H. A novel 1D ladder-like organic-inorganic hybrid compound [Cu(bIz)$_2$]$_2$[{Cu(bIz)$_2$}$_2$Mo$_8$O$_{26}$] (bIz=benzimidazole). Inorg Chem Commun, 2003, 6: 702-705.

[61] Wang J P, Li S Z, Zhao J W, Niu J Y. Synthesis, characterization and crystal structure of a new β-octamolybdate-supported compound: [Ni(H$_2$O)(2,2′-bipy)$_2$]$_2$[Mo$_8$O$_{26}$] · 4H$_2$O · 2CH$_3$COOH. Inorg Chem Commun, 2006, 9: 599-602.

[62] Wang C M, Zeng Q X, Zhang J, Yang G Y. 1D polyoxometalate-based inorganic-organic hybrid derived from β-octamolybdate-synthesis and crystal structure of [Co^{II}(2,2′-bipy)$_2$]$_2$[Mo$_8$O$_{26}$]. Z Anorg Allg Chem, 2005, 631: 838-840.

[63] Chen S M, Lu C Z, Yu Y Q, Zhang Q Z, He X. A new polyoxomolybdate infinite chain complex containing [β-Mo$_8$O$_{26}$]$^{4-}$ linked up by two bonded Ag$^+$ ions: {(Bu$_4$N)$_2$[Ag$_2$Mo$_8$O$_{26}$]}$_n$. Inorg Chem Commun, 2004, 7: 1041-1044.

[64] Shi Z Y, Gu X J, Peng J, Xin Z F. An unprecedented one-dimensional chain constructed from β-octamolybdate clusters and two kinds of silver complex fragments. Eur J Inorg Chem, 2005: 3811-3814.

[65] Abbas H, Pickering A L, Long D L, Kogerler P, Cronin L. Controllable growth of chains and grids from polyoxomolybdate building blocks linked by silver (I) dimers. Chem Eur J, 2005, 11: 1071-1078.

[66] Kitamura A, Ozeki T, Yagasaki A. β-Octamolybdate as a building block. Synthesis and structural characterization of rare earth-molybdate adducts. Inorg Chem, 1997, 36: 4275-4279.

[67] Wu C D, Lu C Z, Lin X, Zhuang H H, Huang J S. Two new β-octamolybdate supported rare earth metal complexes: [NH$_4$]$_2$[{Gd(DMF)$_7$}$_2$(β-Mo$_8$O$_{26}$)][β-Mo$_8$O$_{26}$] and [NH$_4$][La(DMF)$_7$(β-Mo$_8$O$_{26}$)]. Inorg Chem Commun, 2002, 5: 664-666.

[68] Chen S M, Lu C Z, Yu Y Q, Zhang Q Z, He X. Catena-poly [[heptakis (dimethyl-formamide-kppaO) di-μ_4-oxo-tetra-μ_3-oxo-hexadeca-μ_2-oxo-tetraoxolanthanum (III) octamolybdenum (IV)]-μ-sodium (I)]. Acta Cryst, 2004, C60: m549-m550.

[69] Niven M L, Cruywagen J J, Heyns J B B. The first observation of γ-octamolybdate: synthesis, crystal and molecular structure of $[Me_3N(CH_2)_6NMe_3]_2[Mo_8O_{26}] \cdot 2H_2O$. J Chem Soc Dalton Trans, 1991: 2007-2011.

[70] Cui X B, Lu K, Fan Y, Xu J Q, Ye L, Sun Y H, Li Y, Yu H H, Yi Z H. A novel γ-octamolybdate supported transition metal complex $[Cu(imi)_2]_4[\gamma\text{-}Mo_8O_{26}]$. J Mol Struct, 2005, 743: 151-155.

[71] Isobe M, Marumo F, Yamase T, Ikawa T. The crystal structure of hexakis (isopropyl ammonium) dihydrogenoctamolybdate (6−) dihydrate, $(C_3H_{10}N)_6[H_2Mo_8O_{28}] \cdot 2H_2O$. Acta Cryst, 1978, B34: 2728.

[72] Adams R D, Klemperer W G, Liu R S. Synthesis and X-ray structure of a formylated octamolybdate cluster $[(HCO)_2(Mo_8O_{28})]^{6-}$. J C S Chem Comm, 1979: 256-257.

[73] Mccarron E M, Harlow R L. Synthesis and structure of $Na_4[Mo_8O_{24}(OCH_3)_4] \cdot 8CH_3OH$: A novel isopolymolybdate that decomposes with the loss of formaldehyde. J Am Chem Soc, 1983, 105: 6179-6181.

[74] Mccarron E M, Whitney J F, Chase D B. Pyridinium molybdates, synthesis and structure of an octamolybdate containing coordinately bound pyridine: $[(C_5H_5N)_2Mo_8O_{26}]^{4-}$. Inorg Chem, 1984, 23 (21): 3275-3280.

[75] Kamenar B, Korpar-Colig B, Penavic M, Cindric M. Synthese and characterization of octamolybdates containing co-ordinatively bound salicylideneiminato and methioninato (MetO) ligands. Crystal structures of $[NH_3Pr]_2[Mo_8O_{22}(OH)_4(OC_6H_4CH=NPr-2)_2]_2 \cdot 6MeOH$ and $[morph]_4[Mo_8O_{24}(OH)_2(MetO)_2] \cdot 4H_2O$ (morph = morpholine). J Chem Soc Dalton Trans, 1990: 1125-1130.

[76] Martin-Zarza P, Arrieta J M, Munoz-Roca M C, Gili P. Synthesis and characterization of new octamolybdates containing imidazole, 1-methyl or 2-methyl-imidazole co-ordinatively bound to molybdenum. J Chem Soc Dalton Trans, 1993: 1551-1557.

[77] Inoue M, Yamase T. Synthesis and crystal structures of γ-type octamolybdates coordinated by chiral lysines. Bull Chem Soc Jpn, 1995, 68: 3055-3063.

[78] Gili P, Lorenzo-Luis P A, Mederos A, Arrieta J M, Germain G, Castineiras A, Carballo R. Crystal structures of two new heptamolybdates and of a pyrazole incorporating a γ-octamolybdate anion. Inorg Chim Acta, 1999, 295: 106-114.

[79] Laduca R L Jr, Rarig R S Jr, Zapf P J, Zubieta J. Dipyridylamine-ligated octamolybdate clusters aggregated into extended ribbons and layers via hydrogen bonding. Inorg Chim Acta, 1999, 292: 131-136.

[80] Wang S, Lin X, Lu C Z, Wu D M, Yang W B, Zhuang H H. The crystal structure of hexakisammonium diacetyloctamolybdate tetrahydrare. Chin J Struct Chem, 2000, 19 (3): 191-194.

[81] Wu C D, Lu C Z, Chen S M, Zhuang H H, Huang J S. Synthesis and characterization of two new polyoxomolybdate compounds: $[Cu(imi)_2(H_2O)_4][Himi]_2$

[(imi)$_2$Mo$_8$O$_{26}$] and [Himi]$_3$[H$_3$O][SiMo$_{12}$O$_{40}$]·H$_2$O. Polyhedron, 2003, 22: 3091-3095.

[82] Wu C D, Lu C Z, Yang W B, Zhuang H H, Huang J S. Hydrothermal assembly and structural characterization of an organic-inorganic hybrid octamolybdate supported transition metal complex [Cu(imi)$_2$]$_4$[(imi)$_2$Mo$_8$O$_{26}$]·4H$_2$O. J Cluster Sci, 2002, 13 (1): 55-62.

[83] Kang J, Zhang Q Z, Wu C D, Yang W B, Zhan X P, Yu Y Q, Lu C Z. Synthesis and crystal structure of a new imidazole coordinated octamolybdate compound. Chinese J Struct Chem, 2003, 22 (2): 190-194.

[84] Kang J, Zhang Q Z, Wu C D, Yang W B, Zhan X P, Yu Y Q, Lu C Z. Synthesis and crystal structure of tetraimidazolium dinicotinatooctamolybdate. Chinese J Struct Chem, 2003, 22 (1): 84-88.

[85] He X, Ye J W, Xu J N, Fan Y, Wang L, Zhang P, Wang Y. Self-assembly of a 3D supramolecular architecture with nicotinic acid ligands and polyoxomolybdate units. J Mol Struct, 2005, 749: 9-12.

[86] 韩正波, 安海艳, 王力, 栾国有, 王恩波, 韩正学. 有机-无机杂化材料{[Na$_2$(H$_2$O)$_4$]$_2$[γ-Mo$_8$O$_{26}$(Gly-Gly)$_2$]}·4H$_2$O 的合成及晶体结构. 高等学校化学学报, 2003, 24 (9): 1558-1560.

[87] Wu X Y, Liu J H, Zhang Q Z, He X, Chen S M, Lu C Z. A polyoxomolybdate coordinated by glycine ligands: K$_4$[Mo$_8$O$_{26}$(NH$_3$CH$_2$COO)$_2$]·6H$_2$O. Acta Cryst, 2004, E60: m921-m923.

[88] Chen Z F, Liu B, Liang H, Hu R X, Zhou Z Y. Synthesis and crystal structure of [Hamp]$_4$[Mo$_8$O$_{26}$(DMF)$_2$]·2H$_2$O (amp = 2-amino-6-methyl-pyridine). Chinese J Struct Chem, 2004, 23 (1): 19-24.

[89] Sun C Y, Wang E B, An H Y, Xiao D R, Xu L. Synthesis and characterization of a novel organic/inorganic hybrid based on octamolybdates and benzimidazole molecules [Hbenzimi]$_4$[(benzimi)$_2$Mo$_8$O$_{26}$]·2H$_2$O (benzimi = benzimidazole). Transition Metal Chemistry, 2005, 30: 873-878.

[90] Xi R M, Wang B, Isobe K, Nishioka T, Toriumi K, Ozawa Y. Isolation and X-ray crystal structure of a new octamolybdate: [(RhCp*)$_2$(μ_2-SCH$_3$)$_3$]$_4$[Mo$_8$O$_{26}$]·2CH$_3$CN(Cp* = η^5-C$_5$Me$_5$). Inorg Chem, 1994, 33: 833-836.

[91] Hagrman D, Zubieta C, Rose D J, Zubieta J, Haushalter R C. Composite solids constructed from one-dimensional coordination polymer matrices and molybdenum oxide subunits: polyoxomolybdate clusters within [{Cu(4,4'-bpy)}$_4$Mo$_8$O$_{26}$] and [{Ni(H$_2$O)$_2$(4,4'-bpy)}$_2$Mo$_8$O$_{26}$] and one-dimensional oxide chains in [{Cu(4,4'-bpy)}$_4$Mo$_{15}$O$_{47}$]·8H$_2$O. Angew Chem Int Ed Engl, 1997, 36 (8): 873-876.

[92] Rarig R S Jr, Zubieta J. Hydrothermal synthesis and structural characterization of an organic-inorganic hybrid material: (H$_2$tptz)$_2$[δ-Mo$_8$O$_{26}$]·2H$_2$O(tptz = 2,4,6-tripyridyl-

triazine). Inorg Chim Acta, 2001, 312: 188-196.

[93] Xiao D R, An H Y, Wang E B, Xu L. Syntheses and structures of two novel inorganic-organic hybrid octamolybdates: [H_2enMe]$_2$[Mo_8O_{26}] · 2H_2O and [Ni(2,2'-bpy)$_3$]$_2$[δ-Mo_8O_{26}]. J Mol Struct, 2005, 738: 217-225.

[94] Allis D G, Rarig R S, Burkholder E, Zubieta J. A three-dimensional bimetallic oxide constructed from octamolybdate clusters and copper-ligand cation polymer subunits. A comment on the stability of the octamolybdate isomers. J Mol Struct, 2004, 688: 11-31.

[95] Burkholder E, Zubieta J. A two-dimensional bimetallic oxide constructed from ζ-octamolybdate clusters and Ag(Ⅰ)-tpyprz cationic polymer components (tpyprz=tetra-2-pyridylpyrazine). Solid State Sci, 2004, 6: 1421-1428.

[96] Xiao D R, Hou Y, Wang E B, Wang S T, Li Y G, Xu L, Hu C W. Hydrothermal synthesis and characterization of an unprecedented η-type octamolybdate: [{Ni(phen)$_2$}$_2$(Mo_8O_{26})]. Inorg Chim Acta, 2004, 357: 2525-2531.

[97] Allis D G, Burkholder E, Zubieta J. A new octamolybdate: observation of the θ-isomer in [Fe(tpyprz)$_2$]$_2$[Mo_8O_{26}] · 3.7H_2O (tpyprz = tetra-2-pyridylpyrazine). Polyhedron, 2004, 23: 1145-1152.

[98] Zhang D, Zhao J, Zhang Y, Hu X, Li L, Ma P, Wang J, Niu J. Octamolybdate-supported tricarbonyl metal derivatives: [{$H_2Mo_8O_{30}$}{M(CO)$_3$}$_2$]$^{8-}$ (M=Mn^I and Re^I), Dalton Trans, 2013, 42: 2696-2699.

[99] Niu J, Yang L, Zhao J, Ma P, Wang J. Novel octatungstate-supported tricarbonyl metal derivatives: {[$H_2W_8O_{30}$][M(CO)$_3$]$_2$}$^{8-}$ (M = Mn^I and Re^I), Dalton Trans, 2011, 40 (33): 8298-8300.

第9章

九金属氧簇

到目前为止,九金属氧簇是报道最少的一类多金属氧簇。尽管 Yamase 等得到了分子式为 $(NH_4)_6[Mo_9O_{30}] \cdot 5H_2O$ 的化合物[1],但晶体结构研究表明该化合物阴离子只有链状结构。

Kortz 报道的 $Na_7[HW_9O_{33}Ru_2\{(CH_3)_2SO\}_6] \cdot 25.5H_2O$ 为第一个含有九金属氧簇的化合物[2]。该化合物由 cis-$Ru(dmso)_4Cl_2$ 与 $Na_2WO_4 \cdot 2H_2O$ 在 0.5mol/L pH=4.8 的 NaAc 缓冲溶液中 80℃反应 1h 制得,阴离子结构如图 9-1 所示。阴离子通过六个多面体上的多个端基氧原子与两个 $Ru\{(CH_3)_2SO\}_3^{3+}$ 配离子配位。如果去掉阴离子所担载的两个 $Ru\{(CH_3)_2SO\}_3$,九金属氧簇阴离子 $[HW_9O_{33}]^{11-}$ 具有图 9-2 所示的结构。

阴离子中的九个金属原子分为三组,构成三个共边的三金属簇,三组三金属簇之间通过共顶点相连。阴离子中的氧原子可分为三类:①端基氧原子(O_t),共 15 个(含与 Ru 配位的氧原子);②二重桥氧原子(O_b),共 15 个;③三重桥氧原子(O_c),共 3 个。阴离子中的金属原子可分为两类,一类连接一个端基氧原子、四个二重桥氧原子和一个三重桥氧原子(Mo_I),共 3 个;另一类连接两个端基氧原子、三个二重桥氧原子和一个三重桥氧原子(Mo_{II}),共 6 个。每一

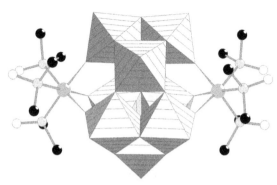

图 9-1　阴离子 $[HW_9O_{33}Ru_2\{(CH_3)_2SO\}_6]^{7-}$ 结构

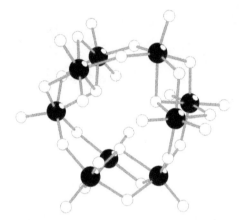

图 9-2　$[HW_9O_{33}Ru_2\{(CH_3)_2SO\}_6]^{7-}$ 中钨核的原子连接图

组三金属氧簇由一个 Mo_I 原子和两个 Mo_{II} 原子构成。每一个 Mo_{II} 通过一个端基氧原子与配离子 $Ru\{(CH_3)_2SO\}_3$ 相连。

到目前为止,还没有分离出具有分立结构的 $[HW_9O_{33}]^{11-}$。

参考文献

[1] Yamase T, Sugeta M, Ishikawa E. Zigzag chain structure of hexaammonium enneamolybdate pentahydrate. Acta Cryst, 1996, C52: 1869-1872.

[2] Bi L, Hussain F, Kortz U, Sadakane M, Dickman M H. A novel isopolytungstate functionalized by ruthenium: $[HW_9O_{33}Ru_2^{II}(dmso)_6]^{7-}$. Chem Commun, 2004: 1420-1421.

第 10 章

十金属氧簇

能生成十金属氧簇的元素比较多，V、Nb、Mo、W 都有十金属氧簇化合物的报道，但不同的元素构成的十金属氧簇在结构上有明显的差别。

10.1 十钒氧簇

由十钒氧簇阴离子构成的化合物是目前十金属氧簇中报道得最多的一类化合物，十钒氧簇阴离子也是自然界唯一存在矿物的同多金属阴离子，至少有三种矿石中存在十钒氧簇阴离子，橙钒钙矿（$Ca_3V_{10}O_{28} \cdot 17H_2O$）、水钒镁石（$K_2Mg_2V_{10}O_{28} \cdot 17H_2O$）和水钒镁钠石（$Na_4MgV_{10}O_{28} \cdot 24H_2O$），人们在很早就开始了对十钒钠氧簇阴离子的研究[1]。

10.1.1 合成

将偏钒酸钠溶液酸化到 pH 为 6.0～2.0 时，橙色溶液中存在的主要物质即为钒氧簇阴离子。继续酸化该溶液，则该溶液中存在如下平衡：

$$V_{10}O_{28}^{6-} + H^+ \rightleftharpoons HV_{10}O_{28}^{5-}$$

$$HV_{10}O_{28}^{5-} + H^+ \rightleftharpoons H_2V_{10}O_{28}^{4-}$$
$$H_2V_{10}O_{28}^{4-} + H^+ \rightleftharpoons H_3V_{10}O_{28}^{3-}$$
$$H_3V_{10}O_{28}^{3-} + H^+ \rightleftharpoons H_4V_{10}O_{28}^{2-}$$

上述表明，$V_{10}O_{28}^{6-}$、$HV_{10}O_{28}^{5-}$ 和 $H_2V_{10}O_{28}^{4-}$ 都可以在溶液中稳定存在。但当溶液中十钒氧簇阴离子主要存在形式 $H_3V_{10}O_{28}^{3-}$ 的浓度大到不稳定的 $H_4V_{10}O_{28}^{2-}$ 出现时，阴离子将会分解[2-5]。

由于在溶液中存在多个平衡，在不同的浓度下，十钒氧簇的质子化程度不同，当阴离子以化合物的形式从溶液析出时，根据析出晶体母液酸度不同，在晶体中，十钒氧簇阴离子也表现出不同的质子化程度。

10.1.2 谱学表征

图 10-1 为十钒氧簇阴离子 $V_{10}O_{28}^{6-}$ 和二质子化的十钒簇阴离子 $H_2V_{10}O_{28}^{4-}$ 的红外光谱[6]。

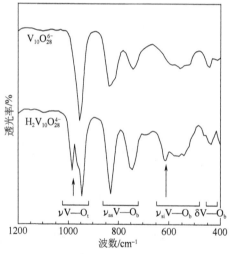

图 10-1 $V_{10}O_{28}^{6-}$ 和 $H_2V_{10}O_{28}^{4-}$ 的红外光谱

阴离子中钒氧双键的振动峰位于 $990\sim900\text{cm}^{-1}$；$640\sim450\text{cm}^{-1}$ 为钒与桥氧相连的键 $V-O_b$ 的伸缩振动；$450\sim400\text{cm}^{-1}$ 的吸收峰为 $V-O_b$ 键的弯曲振动。由图 10-1 中[7-9]可以看出，十钒氧簇阴离子质子化后，$H_2V_{10}O_{28}^{4-}$ 在 985cm^{-1} 和 615cm^{-1} 处的吸收峰由 $V-O_b^{(H)}$ 键振动产生。

10.1.3 晶体结构

尽管十钒氧簇阴离子在自然界以矿物存在，但其结构在 1964 年才被精确测定[10]。后来又有文献对该阴离子的结构进行了研究[5]，阴离子结构如图 10-2 所示。阴离子由 10 个 VO_6 八面体以紧密堆积方式通过共边相连，可以看作首先由六个 VO_6 八面体以 2×3 的方式共边矩形排列，两个共边 VO_6 八面体从上面，另外两个共边 VO_6 八面体从下面通过共用八面体分别同中间的六个八面体相连。阴离子中的 VO_6 八面体可以分为三类：两个 VO_6 八面体（V_{III}）位于 2×3 矩形排列的中间，每一个八面体同相邻的八面体共用七条棱；四个 VO_6 八面体（V_{II}）位于矩形排列的四个角，每一个八面体与相邻的八面体共用四条棱；另外四个八面体位于矩形排列的上部和下部，与相邻的八面体共用五条棱。阴离子中的氧原子分为四类：位于中心与六个钒原子配位的氧原子（O_a），共 2 个；位于阴离子表面，与三个钒原子配位的三重桥氧原子（O_c），共 4 个；位于阴离子表面，与两个钒原子配位的一重桥氧（O_b），共 14 个；只与一个钒原子相连的氧原子（O_d），共 8 个。整个阴离子只有 D_{2h} 对称性。

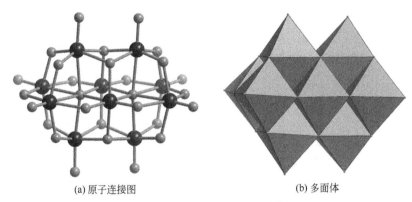

(a) 原子连接图　　　　　　　　(b) 多面体

图 10-2　十钒氧簇阴离子结构

有关十钒氧簇阴离子中质子的位置，有很多文献进行了研究。通过测定酸化十钒氧簇阴离子溶液 ^{17}O 核磁共振，Klemperer 等提出，在质子化的十钒氧簇阴离子中，质子化的位置应该在三重桥氧或二重桥氧上，三重桥氧的质子化能力比二重桥氧的强[11]。通过测定 $H_3V_{10}O_{28}^{3-}$ 的晶体结构，并配合 NMR 等手段，确定 $H_3V_{10}O_{28}^{3-}$ 中的质子 1 个位于三重桥氧，2 个位于二重桥氧上[12]（图 10-3），而且阴离子之间可以通过氢键形成二聚离子（图 10-4）。

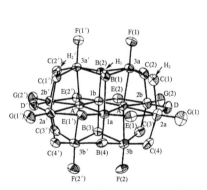 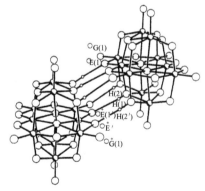

图 10-3　$H_3V_{10}O_{28}^{3-}$ 中质子化的位置　　图 10-4　$H_3V_{10}O_{28}^{3-}$ 通过氢键形成的二聚离子

量子化学计算结果也表明[13]，在 $[H_3V_{10}]^{6-}$ 中，三重桥氧处的碱性最强，最易于结合质子或小的正电荷基团。对于化合物 $[NH_3(C_6H_{13})]_4[H_2V_{10}O_{28}]$ 的键价计算表明，在 $[H_2V_{10}O_{28}]^{4-}$ 中，两个质子分别位于两个三重桥氧上[14]。

通过分析由 $[H_2V_{10}O_{28}]^{4-}$ 和腺苷形成的化合物 [adenosinium]$_4$（$H_2V_{10}O_{28}$）·11H_2O，发现在该化合物的阴离子中，两个质子分别与处于中心对称的两个二重桥氧相连[15]。从现有文献看，阴离子中的质子可位于三重桥氧上，也可与二重桥氧相连，这可能与阳离子作用有关，表 10-1 列出了已报道化合物中阴离子中质子的位置。在有些化合物中，由于缺乏精确的数据支持，质子化的位置难以确定[29,30]。

表 10-1　$[H_xV_{10}O_{28}]^{(6-x)-}$ 中质子位置分布

化合物	原子位置	文献
$[(CH_3)_3CNH_3]_4[H_2V_{10}O_{28}]$	$2O_b$	[6]
$[DPAH]_5[HV_{10}O_{28}]·2H_2O$	$1O_c$	[9]
$[PDAH_2]_2[H_2V_{10}O_{28}]$	$2O_b$	[9]
$[DPAH]_4[H_2V_{10}O_{28}]·2H_2O$	$1O_b,1O_c$	[9]
$[DPAH]_3[H_3V_{10}O_{28}]$	$2O_b,1O_c$	[9]
$[C_7N_2H_{13}]_4[H_2V_{10}O_{28}]$	$2O_b$	[16]
$[N(CH_3)_4]_4[H_2V_{10}O_{28}]·CH_3COOH·2.8H_2O$	$1O_b,1O_c$	[17]
$[^n(Ba)_4N]_2[H_4V_{10}O_{28}]$	$2O_b,2O_c$	[18]
$[C222(H^+)_2]_2[H_2V_{10}O_{28}]·2.5H_2O$	$2O_b$	[19]
$[OthH]_2[H_2V_{10}O_{28}]·8H_2O$	$2O_c$	[20]
$[H_2dpds]_2[H_2V_{10}O_{28}]·10H_2O$	$2O_c$	[21]
$Rb_2[H_4V_{10}O_{28}]·2Gly·Gly·2H_2O$	$2O_b,2O_c$	[22]
$(NH_4)_4[H_4V_{10}O_{28}]·[NTPH_2]_2·4H_2O$	$2O_b,2O_c$	[23]
$[HdaH_2]_2[H_2V_{10}O_{28}]$	$2O_c$	[24]
$[H_2en]_3[H_3V_{10}O_{28}]_2·en·7H_2O$	$2O_b,1O_c$	[25]
$[(C_2H_5)_3NH]_4[H_2V_{10}O_{28}]$	$1O_b,1O_c$	[25]
$[H_4bim][H_2V_{10}O_{28}]·4H_2O$	$2O_b$	[26]
$[NH_4]_2[H_2V_{10}O_{28}]·4(C_8H_{10}N_4O_2)·2H_2O$	$2O_b,2O_c$	[27]
$[C_4N_2S_2H_{14}]_2[H_2V_{10}O_{28}]·4H_2O$	$2O_b$	[28]

注：PDA-1,3-丙二胺；DPA-二丙胺；C222—4,7,13,16,21,24—六氧-1,10-二氮双环 [8.8.8] 二十六烷；dpds—4,4′-二吡啶基二硫；Had-己二胺；Bim—2,2-双（4,5-二甲基咪唑）。

10.1.4　相关化合物

十钒氧簇阴离子作为一种稳定的阴离子，可以与各种类型的阳离子结合形成化合物。表 10-2 列出了部分已报道的由十钒氧簇阴离子形成的具有分立结构的化合物及其简要合成方法。

表 10-2　报道的部分十钒氧簇化合物

化合物	合成	文献
$[(CH_3)_3CNH_3]_6[V_{10}O_{28}] \cdot 8H_2O$	1g V_2O_5,30mL 水,0.75mL t-丁胺,回流,搅拌,pH=5.5	[6]
$[(CH_3)_3CNH_3]_4[H_2V_{10}O_{28}]$	2.13g V_2O_5,1mL t-丁胺,100mL 水,回流,pH=3.5	[6]
$[PDAH_2]_3[V_{10}O_{28}] \cdot 5H_2O$	0.55g V_2O_5,1mL 1,3-丙二胺,水 100mL,pH=6.0	[9]
$[PDAH_2]_2[H_2V_{10}O_{28}]$	1.1g V_2O_5,1mL 1,3-丙二胺,100mL 水,pH=3.0	[9]
$[DPAH]_5[HV_{10}O_{28}] \cdot 2H_2O$	0.5g V_2O_5,1.5mL 二丙胺,40mL H_2O,pH=5.5	[9]
$[DPAH]_4[H_2V_{10}O_{28}] \cdot 2H_2O$	0.5g V_2O_5,1.5mL 二丙胺,40mL H_2O,pH=2.7	[9]
$[DPAH]_3[H_3V_{10}O_{28}]$	0.67g V_2O_5,1.5mL 二丙胺,40mL H_2O,pH=2.7	[9]
$[(C_6H_{13})NH_3][H_2V_{10}O_{28}]$	0.3g $CuCl_2$,2.0g V_2O_5,40mL 水,1.8mL 环己胺,加热,搅拌 30min	[14]
$[Adem]_4[H_2V_{10}O_{28}] \cdot 11H_2O$	$VOSO_4$ 水溶液,等物质的量腺苷,混合振荡	[15]
$[C_7N_2H_{13}]_4[H_2V_{10}O_{28}] \cdot 6H_2O$	0.08mL 乙酰丙酮,1mL 0.4mol/L 乙二胺水溶液,2mL 0.2mol/L $VOSO_4$	[16]
$[N(CH_3)_4]_4[H_2V_{10}O_{28}] \cdot CH_3COOH \cdot 2.8H_2O$	V_2O_5,$Al(NO_3)$,25% 四甲基胺水溶液按 1:2:4 摩尔比,30mL 3mol/L 醋酸,pH=2.94,水热 473k,4d	[17]
$(^nBu_4N)_2[H_4V_{10}O_{28}]$	170.7mg $(^nBu_4N)VO_3$,30mL 甲醇,3～5mL 乙酸	[18]
$[HdaH_2]_2[H_2V_{10}O_{28}]$	0.45g V_2O_5,0.3g 乙二胺溶于水稀释至 100mL,pH=3.1	[24]
$[HdaH_2]_2[H_2V_{10}O_{28}] \cdot 2H_2O$	0.45g V_2O_5,0.3g 乙二胺溶于水稀释至 100mL,pH=3.5	[24]
$[H_2en]_2K_2[V_{10}O_{28}] \cdot 4H_2O$	0.78g $Na_2H_2(SO_3\text{-}Sal)_2$en 溶于 5mL 水,30mL 0.012mol/L 钒酸盐溶液,用 HCl 调至 pH=5,低温 5℃,放置溶液 2d	[25]
$[H_2en]_3[H_3V_{10}O_{28}]$en $\cdot 7H_2O$	0.78g $Na_2H_2(SO_3\text{-}Sal)_2$en 溶于 5mL 水,30mL 0.012mol/L 钒酸盐溶液,用 HCl 调至 pH=4,低温 5℃,放置溶液 2d	[25]
$[(C_2H_5)_3NH]_4[H_2V_{10}O_{28}]$	0.14g D,L-二氨基丙酸,0.23mL 水杨醛,0.2531g $VOSO_4$,用三乙胺调至 pH=5	[25]

续表

化合物	合成	文献
$[H_4bim]_2[H_2V_{10}O_{28}] \cdot 4H_2O$	1.085g $VOSO_4 \cdot nH_2O$，100mL 水，0.42g LiOH，0.95g Hbim 溶于 350mL 乙醇，混合液搅拌 24h	[26]
$[NH_4]_2(C_8H_{10}N_4O_2)_4[H_4V_{10}O_{28}] \cdot 2H_2O$	0.50g NH_4VO_3，20mL 水，pH=4.5，1.71mmol 咖啡因	[27]
$[C_4N_2S_2H_{14}]_2[H_2V_{10}O_{28}] \cdot 4H_2O$	0.4542g V_2O_5，1.2g NaOH，0.385g 巯基乙胺，40mL 水，pH=3.0	[28]
$[C_2N_2H_8O_2]_3H_6V_{10}O_{28}$	6.0g NaOH，50mL 水，7.5g C_2H_5COOH，7.0g $NH_2OH \cdot HCl$，3.8g $NaBH_4$，2mL 30% H_2O_2，pH=6.5～7.0，30mL $VOSO_3 \cdot 3H_2O$ (11g)水溶液	[29]
$\{[Na_2(H_2O)_4]_2\}(\mu-H_2O)_2[H_3O]_2[H_2V_{10}O_{28}]$	等量硫酸氧钒、香草醛、胱氨酸于醋酸-醋酸钠缓冲溶液中放置	[30]
$[C211H_2]_2[H_3O]_2[V_{10}O_{28}] \cdot 7H_2O$	58mg C211，20mL 10mmol/L 十钒酸盐溶液，pH=5.5，4℃放置	[30]
$[C23H_2]_2[H_2V_{10}O_{28}] \cdot 6H_2O$	6.9mg C23，20mL 10mmol/L 十钒酸盐溶液，pH=5.5，4℃放置	[30]
$[Me_2NH_2]_6[V_{10}O_{28}] \cdot H_2O$	151mg 鸟嘌呤，50mL DMF，157mg VO_3，氮气，80℃，2h	[31]
$[CN_3H_6]_6[V_{10}O_{28}]$	4.5g $NaVO_3$，50mL 水，5.0g 盐酸胍，用 3mol/L 硝酸调至 pH=5～6	[32]
$(NH_4)_6[V_{10}O_{28}] \cdot 2(Gly-Gly) \cdot 4H_2O$	1.00g NH_4VO_3，50mL 水，pH=5.80，1.46g 盐酸二肽，室温搅拌 12h	[33]
$[Mn(H_2O)_6]_2[(CH_3)_4N]_2[V_{10}O_{28}] \cdot 2H_2O$	100mg V_2O_5，126mg $MnCO_3$，84mg $(CH_3)_4NBr$，0.5mL H_2O，100℃，反应 7d	[34]
$[Eu(CH_2O)_8]_2[V_{10}O_{28}] \cdot 8H_2O$	0.27g 偏钒酸钠，20mL 水，硝酸酸化 pH=4.1，0.15g $EuCl_3 \cdot 6H_2O$，溶于 5mL 水，室温反应	[35]
$[La(H_2O)_9]_2[V_{10}O_{28}] \cdot 8H_2O$	1mmol $Na_6V_{10}O_{28} \cdot 12H_2O$，2mmol $La(NO_3)_3$，100 mL 水，pH=4.0，277K 放置	[36]
$(NH_4)_2[M(dod)(H_2O)_4]_2[V_{10}O_{28}] \cdot 6H_2O$ (M=Mn,Zn)	0.090g V_2O_5，0.243g dod $\cdot$ 2HCl，0.075g $MnCl_2$（或 0.068g $ZnCl_2$），20mL 水，70℃搅拌 30min，滴加 5mL 0.65g 硫酸肼水溶液，用氨水调至 pH=6	[37]
$(C_2H_{10}N_2)_3[V_{10}O_{28}] \cdot 2H_2O$	300mg $VO(VO_4) \cdot 0.5(N_2C_2H_{10})$ 溶于 50mL H_2O_2	[38]
$[Li(H_2O)_4]_2[N(CH_3)_4]_4[V_{10}O_{28}] \cdot 4H_2O$	V_2O_5、氯化四甲基胺、LiOH 摩尔比为 1∶2∶1，硝酸酸化 pH=3.7，473K 微波 2h	[39]

续表

化合物	合成	文献
$Ca_2(H_3O)_2[V_{10}O_{28}] \cdot 16H_2O$	M_oO_3,$Ca(OH)_2$,NH_4NO_3 溶于热水,用盐酸调至 pH=6,加入 $CaCl_2 \cdot H_2O$	[40]
$(NH_4)_4Na_2[V_{10}O_{28}] \cdot 10H_2O$	250mg NH_4VO_3,50mg NaH_2PO_4,50mL H_2O,室温挥发	[41]
$Cs_4[Na_2(H_2O)_{10}][V_{10}O_{28}]$	0.221g $NaVO_3 \cdot H_2O$,2mL 水,15mL 0.5% HF 水溶液,0.217g CsF 的 2mL 水溶液	[42]
$(NH_4)_4[Na_2(H_2O)_{10}][V_{10}O_{28}]$	1.0g $NaMoO_4$,0.2g NH_4VO_3,1.9g $N_2H_4 \cdot 2HCl$,0.6g CH_3COONH_4,10mL 水,150℃反应 60h	[43]
$(NH_4)_4Li_2[V_{10}O_{28}] \cdot 10H_2O$	NH_4VO_3、$LiNO_3$ 摩尔比 2:1,pH=5,343K 反应	[44]
$[Habo]_6[V_{10}O_{28}] \cdot 2C_3H_7OH \cdot 2H_2O$	273mg VO_2(acac),40mL 乙腈,89mg abo,室温搅拌 6h	[45]
$[C_3N_2H_{12}]_3[V_{10}O_{28}] \cdot 5.5H_2O$	0.095g $VOC_2O_4 \cdot 2H_2O$ 溶于 12mL 95% 甲醇,0.037g 1,3-丙二胺溶于 5mL 甲醇,333K 搅拌 6h	[46]
$[C_4H_{14}N_2]_3[V_{10}O_{28}] \cdot 6H_2O$	0.09g 1,4-丁二胺溶于 30mL 水,加入 0.23g V_2O_5,40mL 水,343~353K 保温 0.5h	[47]
$(C_3H_{12}N_2)_2[Cu(H_2O)_6][V_{10}O_{28}] \cdot 7H_2O$	0.064g NH_4VO_3 溶于醋酸水溶液(pH=1~2),加入 1mL $Cu(CH_3COO)_2$ 水溶液(0.01mol/L)和 1mL 1,3 丙二胺水溶液(0.1mol/L)	[48]
$(NH_4)_2[Ni(H_2O)_6]_2[V_{10}O_{28}] \cdot 4H_2O$	NH_4NO_3、$Ni(OAc)_2 \cdot 4H_2O$ 摩尔比 5:1,pH=5,398K 水热反应 10h	[49]
$[C_4H_{14}N_2]_3[V_{10}O_{28}] \cdot 5H_2O$	0.191g $VOC_2O_4 \cdot 2H_2O$ 溶于 25mL 95% 甲酸,加入 5mL N,N-二甲基乙二胺(0.088g)甲醇溶液中,333K 反应 10h	[50]
$[C_5H_{16}N_2]_3[V_{10}O_{28}] \cdot 6H_2O$	0.095g $VOC_2O_4 \cdot 2H_2O$ 溶于 12mL 95% 甲酸,加入 0.051g N,N-二甲基丙二胺(1,3 二胺)溶于 5mL 甲醇溶液中,333K 反应 6h	[51]
$[Cu(pyr)(H_2O)_4]_2(H_3O)_2[V_{10}O_{28}] \cdot 13.5H_2O$	0.11g H_4pmida,0.10g $VOSO_4 \cdot 5H_2O$,0.125g $Cu(OAc)_2 \cdot H_2O$,0.04g 吡嗪,0.1g NaOH,8mL 水和 5mL 乙醇,回流 5h	[52]
$[Ni(pyr)(H_2O)_4]_2(H_3O)_2[V_{10}O_{28}] \cdot 9.5H_2O$	0.11g H_4pmida,0.155g $VOSO_4 \cdot 5H_2O$,0.155g $Ni(OAc)_2 \cdot H_2O$,0.04g 吡嗪,0.1g NaOH,8mL 水和 5mL 乙醇,回流 5h	[52]
$[Zn(H_2O)_{14}(V_{10}O_{28})] \cdot H_2$ppz	0.11g H_4pmida,0.11g $VOSO_4 \cdot 5H_2O$,0.110g $ZnC_4H_4O_6 \cdot 2H_2O$,0.04g 吡嗪,0.1g NaOH,8mL 水和 5mL 乙醇,回流 5h	[52]

十钒氧簇阴离子除了可以和各类阳离子形成分立结构的化合物外，还可以与碱金属、碱土金属以及过渡金属配位相连，形成不同结构类型的化合物。

向 70~72℃的偏钒酸钾 KVO_3（6mmol 溶于 25mL 水中）溶液中加入 $NaMoO_4 \cdot 2H_2O$(1mmol) 和 $CoCl_2 \cdot 6H_2O$(3mmol 溶于 4mL 水中)，用 H_2SO_4 调节 pH 为 3.5，70~72℃ 保温 12~16h，可得到单元组成为 $[K(H_2O)_5Co(H_2O)_3]_2[V_{10}O_{28}]$ 的化合物。用等量的 $NiCl_2 \cdot 6H_2O$ 代替 $CoCl_2 \cdot 6H_2O$ 可以得到同构的 $[K(H_2O)_5Ni(H_2O)_3]_2[V_{10}O_{28}]$[53]，该化合物的结构单元如图 10-5 所示。

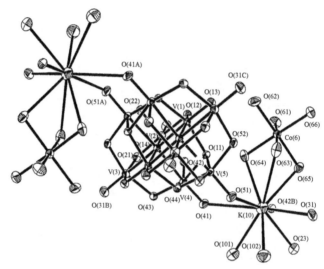

图 10-5　$[K(H_2O)_5Co(H_2O)_3]_2[V_{10}O_{28}]$ 结构单元

十钒氧簇阴离子通过 4 个端基氧原子分别与两个钾离子相连，每个钾离子除通过三个桥联水分子与 Co(Ni) 离子相连外，还与另一个阴离子的两个端基氧原子相连，构成一维链状结构（图 10-6），钾离子还与相邻链上十钒氧簇的一个端基氧原子相连，构成二维层状结构。

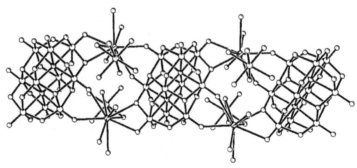

图 10-6　$[V_{10}O_{28}]^{6-}$ 与阳离子构成的一维链状结构

第10章 十金属氧簇

0.4652g V_2O_5 和 5.0mL 6mol/L NaOH 在 20mL 水中冰浴搅拌，加入 0.5mL 30% H_2O_2 并用 6mol/L HNO_3 调 pH 为 4.10，加水至体积为 25mL，加入 5mL 1mol/L $Ni(NO_3)_2 \cdot 6H_2O$ 水溶液，再加入 7.5mL 乙醇，可得到结构单元为 $[Ni(H_2O)_6]_2[Na(H_2O)_3]_2[V_{10}O_{28}] \cdot 4H_2O$ 的化合物[54]。在化合物的结构单元中，十钒氧簇阴离子通过一个端基氧原子与钠离子配位，钠离子通过两个桥联水分子与 Ni^{2+} 相连，Ni^{2+} 通过另外两个桥联水分子与一个钠离子配位，该钠离子又与另一个十钒氧簇阴离子的一个端基氧原子相连，构成一维链状结构。图 10-7 为化合物的结构单元。

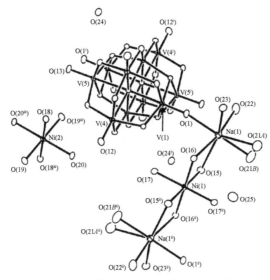

图 10-7 化合物 $[Ni(H_2O)_6]_2[Na(H_2O)_3]_2[V_{10}O_{28}] \cdot 4H_2O$ 结构单元

将 1.22g 偏钒酸钠溶于 100mL 热水中，用 CH_3COOH 调 pH 值为 3.7，加入溶有 0.65g $Mg(CH_3COO)_2 \cdot 4H_2O$ 的 10mL 水溶液，用丙酮扩散法生长晶体，可得到结构单元组成为 $Mg_2Na_2V_{10}O_{28} \cdot 20H_2O$ 的化合物[55]。在该化合物结构单元中，Mg^{2+} 分别与六个水分子配位，形成 $Mg(H_2O)_6^{2+}$ 配阳离子，两个 Na^+ 通过两个水分子桥联，每个钠离子分别与一个十钒氧簇的桥氧和两个十钒氧簇的端氧相连，构成三维网状结构。3.64g V_2O_5 溶于 H_2O_2 水溶液（50mL 30% H_2O_2 稀释至 400mL），加入 2.68g 溶于 10mL 水 $Mg(CH_3COO)_2 \cdot 4H_2O$ 水溶液，加热蒸发至 100mL，用丙酮扩散生长晶体，得到分子组成为 $Mg_3V_{10}O_{28} \cdot 28H_2O$ 的化合物[55]。在该化合物中，三个 Mg^{2+} 分别与六个水分子配位，配离子 $Mg(H_2O)_6^{2+}$ 和十钒氧簇阴离子间通过静电引力结合。0.061g $NaVO_3$、0.065g 硫酸肼和 0.121g dod·2HCl 在 20mL 50℃ 水中搅拌 10min，可得到结构单元组成为 $[Na_6(H_2O)_{16}(dod)_4][V_{10}O_{28}]$ 的化合物[56]。十钒氧簇

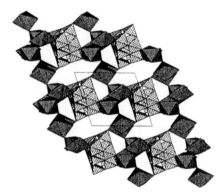

图 10-8 化合物 [Na$_6$(H$_2$O)$_{16}$(dod)$_4$][V$_{10}$O$_{28}$] 构成的三维网状结构

阴离子通过六个端基氧原子分别与四个钠离子相连，构成三维网状结构（图 10-8）。

NH$_4$VO$_3$、LiOH 和 NaOH 按 1∶0.4∶0.2 的摩尔比在水溶液中酸化至 pH=5，水热 398K 反应 10h，得到结构单元为 [LiNa$_2$(H$_2$O)$_9$]$_2$[V$_{10}$O$_{28}$] 的化合物[57]。在该化合物中，一个十钒氧簇阴离子与四个 [LiNa$_2$(H$_2$O)$_9$]$^{3+}$ 基团通过位于矩形排列的 VO$_6$ 八面体四个角位八面体的端基氧相连，每一个 [LiNa$_2$(H$_2$O)$_9$]$^{3+}$ 基团同时与两个十钒氧簇阴离子相连，构成如图 10-9 所示的一维链状结构。

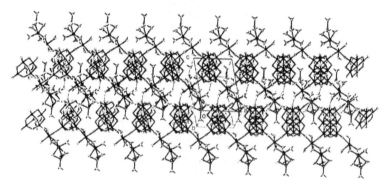

图 10-9 化合物 [LiNa$_2$(H$_2$O)$_9$]$_2$[V$_{10}$O$_{28}$] 构成的一维链状结构

1.48g V$_2$O$_5$ 溶于 200mL H$_2$O$_2$ 水溶液（20mL 30% H$_2$O$_2$ 稀释成），剧烈搅拌 12h，加入 20mL 含 1.28g Cu(CH$_3$COO)$_2$·H$_2$O 的水溶液，用 CH$_3$COOH 调 pH=2.98，由丙酮气相扩散法生长晶体，可得到结构单元组成为 Cu$_3$V$_{10}$O$_{28}$·24H$_2$O 的化合物[58]。在化合物中，其中两个 Cu^{2+} 分别与六个水分子结合，形成 [Cu(H$_2$O)$_6$]$^{2+}$ 配离子；另一个铜离子在与四个水分子配位的同时，还与十钒氧簇阴离子上的两个端基原子配位。每一个十钒氧簇阴离子通过两个矩形排列的 VO$_6$ 八面体的端基氧原子与两个 Cu^{2+} 配位，构成一维链状结构（图 10-10）。

向含有 0.78g NaVO$_3$ 的 50mL 水溶液中加入含 0.512g Cu(CH$_3$COO)$_2$·H$_2$O 的水溶液 20mL，用 CH$_3$COOH 调 pH 值到 3.39，用丙酮气相扩散法生长晶体，可得到结构单元组成为 CuNa$_4$V$_{10}$O$_{28}$·23H$_2$O 的化合物[58]。在化合物中，十钒氧簇阴离子通过位于矩形排列两个对角位 VO$_6$ 八面体的端基氧原子与两个钠离子配位（图 10-11），每一个钠离子再与另一个十钒氧簇阴离子的端基

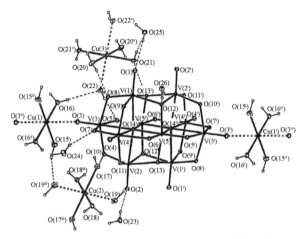

图 10-10 化合物 $Cu_3V_{10}O_{28} \cdot 24H_2O$ 结构单元

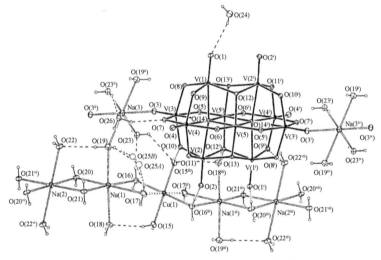

图 10-11 化合物 $CuNa_4V_{10}O_{28} \cdot 23H_2O$ 结构单元

氧原子相连构成一维链状结构；另外两个钠离子和铜离子之间通过两个桥联水分子两两相连构成一维链，两个一维链之间通过与钠离子相连的桥联水分子交叉相连，整个化合物构成三维网状结构。搅拌下将 0.59g MgO 加入由 0.60g NH_4VO_3 溶于 20mL 稀盐酸（37％HCl 和水体积比 1∶4，pH 约为 1）形成的溶液中，再加入 2.40g NaOH 固体，反应体系最终 pH 为 6，用甲醇蒸气扩散法生长单晶，可得到分子组成为 $[Na_4(H_2O)_{14}][Mg(H_2O)_6][V_{10}O_{28}] \cdot 3H_2O$ 的化合物[59]。在化合物中，十钒氧簇阴离子通过两个端基原子与两个钠离子配位，两个钠离子又通过两个桥联水分子分别与两个镁离子相连，进而构成二维层状结构（图 10-12）。

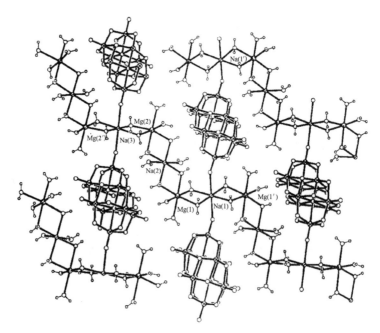

图 10-12 化合物 [Na$_4$(H$_2$O)$_{14}$][Mg(H$_2$O)$_6$][V$_{10}$O$_{28}$]·3H$_2$O 构成的二维层状结构

由硝酸锌、氢氧化铜和五氧化二钒按照 1∶1∶2 的摩尔比加入水中，混合物搅拌加热至沸腾 1h，可得到分子组成为 [Zn(H$_2$O)$_6$][Zn$_2$V$_{10}$O$_{28}$(H$_2$O)$_{10}$]·6H$_2$O 的化合物[60]。化合物分子由 [Zn(H$_2$O)$_6$]$^{2+}$ 和 [Zn$_2$V$_{10}$O$_{28}$(H$_2$O)$_{10}$]$^{2-}$ 及 6 个结晶水组成，阴、阳离子之间通过静电引力相互作用，在 [Zn$_2$V$_{10}$O$_{28}$(H$_2$O)$_{10}$]$^{2-}$ 中，十钒氧簇阴离子通过两个端基氧原子分别与配离子 [Zn(H$_2$O)$_5$]$^{2+}$ 相连（图 10-13）。

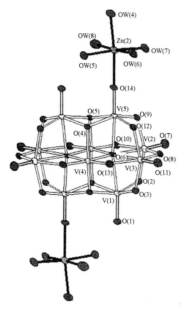

图 10-13 [Zn$_2$V$_{10}$O$_{28}$(H$_2$O)$_{10}$]$^{2-}$ 结构

由 K$_2$[H$_4$V$_{10}$O$_{28}$]·7H$_2$O、Cu(NO$_3$)$_2$·6H$_2$O、2,2′-联吡啶和水（9.0mL）按 1∶1∶2∶1000 的摩尔比混合搅拌 1h，用 HNO$_3$ 调 pH 值到 4.5，160℃反应 4d，可得到分子组成为 [Cu(2,2′-bpy)$_2$][H$_2$V$_{10}$O$_{28}$]·(2,2′-bpy)·H$_2$O 的化合物[61]。在该化合物中，十钒氧簇阴离子通过两个桥氧原子分别与两个 Cu^{2+} 配位，同时，每一个 Cu^{2+} 还与两个 2,2′-联吡啶分子配位，形成担载结构的分子（图 10-14）。

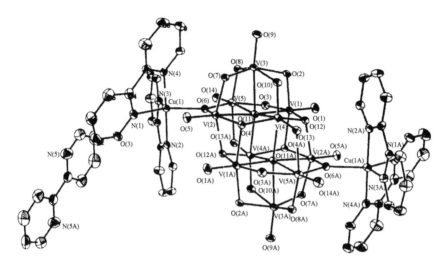

图 10-14 化合物 [Cu(2,2'-bpy)$_2$][H$_2$V$_{10}$O$_{28}$]·(2,2'-bpy)·H$_2$O 分子结构

由 V$_2$O$_5$、NH$_4$VO$_3$、1,1,1-三（羟甲基）丙烷、氯化铵和水在 150℃下反应，得到分子组成为 (NH$_4$)$_4$[V$_{10}$O$_{16}${CH$_3$CH$_2$C(CH$_2$O)$_3$}$_4$]·4H$_2$O 的化合物[62]。在该化合物的阴离子中，四个 1,1,1-三（羟甲基）丙胺中的羟基参与了阴离子的构成。12个羟基中的氧原子替代了 V$_{10}$O$_{28}^{6-}$ 中的 10 个二重、2 个三重桥氧原子，结构如图 10-15 所示。改变反应条件，可以得到分子组成为 (Et$_4$N)[V$_{10}$O$_{13}${CH$_3$CH$_2$C(CH$_2$O)$_3$}$_5$] 和 (Me$_4$N)$_{2/3}$(Et$_4$N)$_{1/3}$[V$_{10}$O$_{13}${CH$_3$C

图 10-15 [V$_{10}$O$_{16}${CH$_3$CH$_2$C(CH$_2$O)$_3$}$_4$]$^{4-}$ 原子连接图

(CH$_2$O)$_3$}$_5$]·0.69H$_2$O 的化合物，在该化合物的阴离子中，5 个 1,1,1-三（羟甲基）丙胺中 15 个羟基中的氧原子替代了 V$_{10}$O$_{28}^{6-}$ 中的 12 个二重桥氧和 3 个三重桥氧原子。

由 V$_2$O$_3$、V$_2$O$_5$、NaVO$_3$、季戊四醇、盐酸三甲胺和水在 150℃下水热反应，可以得到分子组成为 (Me$_3$NH)$_2$[V$_{10}$O$_{14}$(OH)$_2${(OCH$_2$)$_3$CCH$_2$OH}$_4$]·H$_2$O 的化合物。键价计算表明，在该化合物阴离子中，所有的钒均为+4 价，并且其中的两个二重桥氧结合了质子[63]。由 V$_2$O$_3$、V$_2$O$_5$、1,1,1-三（羟甲基）乙烷在水热条件下分别与偏钒酸钠、氯化钠和溴化亚铜或偏钒酸钾、氯化钾和氯化锰反应可分别得到组成为 Na$_2$[V$_{10}$O$_{16}${(OCH$_2$)$_3$CCH$_2$CH$_3$}$_4$] 和 K$_2$

[V₁₀O₁₆{(OCH₂)₃CCH₂CH₃}₄]·2H₂O 的化合物。由 V_2O_3、V_2O_5、1,1,1-三（羟甲基）乙烷、水合硫酸氧钒和 40% 的四丁基氢氧化胺在 150℃ 水热条件下反应，得到分子组成为 [(C₄H₉)₄N]₂[V₁₀O₁₆{(OCH₂)₃CCH₃}₄] 的化合物[63]。

$[V_{10}O_{28}]^{6-}$ 中的氧被多元醇中的羟基氧取代以后，阴离子红外光谱特征振动峰有明显变化，取代后的二重桥氧（V—O—V）特征吸收峰与 $[V_{10}O_{28}]^{6-}$ 特征吸收峰相比发生红移，端氧特征峰则发生蓝移。表 10-3 列出了含有 $[V_{10}O_{28}]$ 核的阴离子红外光谱数据。

表 10-3 含 $[V_{10}O_{28}]$ 核的化合物红外光谱数据   单位：cm^{-1}

化合物	$\nu(V-O_t)$	$\nu(V-O-V)$	参考文献
(NR₄)₃[H₃V₁₀^{V}O₂₈]	968,940	840,803,770	[12]
(NH₄)₄[V₁₀O₁₆EtC{(OCH₂)₃}₄]·4H₂O	972,942	840,773	[62]
(Et₄N)₄[V₁₀^{IV}O₁₃{(OCH₂)₃CCH₂CH₃}₅]	968	777	[64]
(Me₃NH)₂[V₁₀^{IV}O₁₄(OH)₂{(OCH₂)₃CCH₂OH}₄]	974	838	[63]
Na₂[V₈^{IV}V₂^{V}O₁₆{(OCH₂)₃CCH₂CH₃}₄]	998,987,977,964,940	847,775	[63]
K₂[V₈^{IV}V₂^{V}O₁₆{(OCH₂)₃CCH₂CH₃}₄]·2H₂O	995,979,970,961,943	850,780	[63]
(TBA)₂[V₈^{IV}V₂^{V}O₁₆{(OCH₂)₃CCH₃}₄]	983,970,950	838,780	[63]

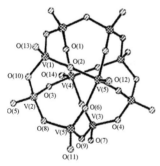

图 10-16 $[V_2^{IV}V_8^VO_{26}]^{4-}$ 结构

含有十个钒原子的金属氧簇阴离子，除广泛报道的 $[V_{10}O_{28}]^{6-}$ 外，还存在组成和结构与此不同的物种。

将 V_2O_5、$Ni(OAc)_2·4H_2O$、1,10-邻菲啰啉和水按一定比例混合在 150℃ 下水热反应，可得到分子组成为 [Ni(1,10-phen)₃]₂·[V₂^{IV}V₈^{V}O₂₆] 的化合物[65]。其阴离子结构如图 10-16 所示。

在 $[V_2^{IV}V_8^VO_{26}]^{4-}$ 中，8 个 VO₄ 四面体通过共顶点相连形成一个闭合的八元环，两个 VO₅ 四方锥通过共用氧原子从环的两侧与八元环上的钒原子交替相连，进而构成组成为 $[V_{10}O_{26}]^{2-}$ 的十钒氧簇阴离子。在该阴离子中，八个四面体中的钒为 +5 价，而在两个 {VO₅} 四方锥中的钒为 +4 价。

10.2 十铌氧簇

10.2.1 合成

1977 年，Morosin 等人首次报道了含十铌氧簇阴离子 $Nb_{10}O_{28}^{6-}$ 化合物的制备[66]。由 178.2g 五乙氧基铌与 127.5g 四甲基氢氧化胺溶液混合，用 200mL 甲醇和 25mL 水稀释，溶液的一半覆盖放置，另一半加入 5mL5% 的氢氧化钠甲醇溶液。24h 以后，可分别得到分子组成为 $[(CH_3)_4N]_6Nb_{10}O_{28} \cdot 6H_2O$ 和 $[(CH_3)_4N]_4Na_2Nb_{10}O_{28} \cdot 8H_2O \cdot 1/2CH_3OH$ 的晶体。制备过程如下：

$$[(CH_3)_4N]_6Nb_{10}O_{28} \cdot 6H_2O \text{ 晶体} \xleftarrow[H_2O]{(CH_3)_4NOH} \boxed{Nb(OC_2H_5)_5 CH_3OH} \xrightarrow[NaOH,H_2O]{(CH_3)_4NOH}$$
$$[(CH_3)_4N]_6Na_2Nb_{10}O_{28} \cdot 8H_2O \cdot 1/2CH_3OH \text{ 晶体}$$

由水合五氧化二铌 $Nb_2O_5 \cdot nH_2O$ 或 $K_7HNb_6O_{19}$ 在水热条件下反应，也可得含十铌氧簇阴离子的化合物[67,68]。

10.2.2 谱学表征

$\{[Zn(2,2'-bpy)_2]_3[Nb_{10}O_{28}] \cdot 3H_2O\}_n$ 和 $\{[Co(2,2'-bpy)_2]_3[Nb_{10}O_{28}] \cdot 1.5H_2O\}_n$ 红外光谱特征吸收峰出现在 1000~400cm^{-1} [67]（图 10-17）。位于 916cm^{-1}，886cm^{-1} 和 859cm^{-1} 的吸收峰为端基氧 $Nb—O_t$ 的特征振动，800~400cm^{-1} 之间的为桥氧 $Nb—O_b—Nb$ 的特征振动。

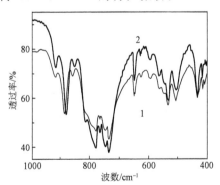

图 10-17 化合物 $\{[Zn(2,2'-bpy)_2]_3[Nb_{10}O_{28}] \cdot 3H_2O\}_n$
和 $\{[Co(2,2'-bpy)_2]_3[Nb_{10}O_{28}] \cdot 1.5H_2O\}_n$ 红外光谱
1—$\{[Zn(2,2'-bpy)_2]_3[Nb_{10}O_{28}] \cdot 3H_2O\}_n$；2—$\{[Co(2,2'-bpy)_2]_3[Nb_{10}O_{28}] \cdot 1.5H_2O\}_n$

10.2.3 晶体结构

X 射线单晶衍射结构分析表明，十铌氧簇阴离子的结构与十钒氧簇阴离子

图 10-18 $[V_{10}O_{28}]^{6-}$ 结构

$[V_{10}O_{28}]^{6-}$ 结构相似。阴离子由 10 个铌原子和 28 个氧原子通过 NbO_6 八面体紧密堆积而成，6 个 NbO_6 八面体以 2×3 的方式通过共边矩形排列，两个共边 NbO_6 八面体通过共边于矩形上面与矩形上的八面体相连，另外两个从下面通过共边与矩形相连（图 10-18）。根据配位环境的不同，阴离子中的氧原子可分为四类：①端基氧原子 O_d，只与一个铌原子相连，共 8 个；②二重桥氧原子 O_c，与两个铌原子相连，共 14 个；③三重桥氧原子 O_b，与三个铌原子相连，共 4 个；④位于中心的六重桥氧原子 O_a，共 2 个。阴离子中 Nb—O 键键长在 1.73~2.55Å，表明阴离子中 NbO_6 八面体畸变严重。

10.2.4 相关化合物

目前报道的与十铌氧簇相关的化合物不多。由 $Zn(NO_3)_2/Co(NO_3)_2$、2,2'-联吡啶、K_7HNbO_{19}、$Na_2S_2O_3$ 和水，在水热条件下反应可生成分子组成为 $\{[Zn(2,2'-bpy)_2]_3[Nb_{10}O_{28}] \cdot 3H_2O\}_n$ 和 $\{[Co(2,2'-bpy)_2]_3[Nb_{10}O_{28}] \cdot 1.5H_2O\}_n$ 的化合物。晶体结构分析表明，化合物分子呈一维链状结构，每一个十铌氧簇阴离子与 4 个 $[M(2,2'-bpy)_2]^{2+}$ 配离子基团相连，2 个 $[M(2,2'-bpy)_2]^{2+}$ 配离子与同一个十铌氧簇阴离子的 2 个端氧相连，另 2 个 $[M(2,2'-bpy)_2]^{2+}$ 配离子分别与 2 个十铌氧簇阴离子上的 2 个端基氧原子相连，构成一维链状结构[67]（图 10-19）。

图 10-19 $\{[Zn(2,2'-bpy)_2]_3[Nb_{10}O_{28}] \cdot 3H_2O\}_n$ 和 $\{[Co(2,2'-bpy)_2]_3[Nb_{10}O_{28}] \cdot 1.5H_2O\}_n$ 的化合物构成的一维链状结构

向 $TBA_4[H_4Nb_6O_{19}]$ 的四氢呋喃溶液中通入 NO 气后过滤，向滤液中加入乙酸乙酯。可析出组成为 $TBA_8[Nb_{20}O_{54}] \cdot H_2O$ 的晶体[69]，结构分析表明，该化合物阴离子 $[Nb_{20}O_{54}]^{8-}$ 为十铌氧簇阴离子 $Nb_{10}O_{28}^{6-}$ 的二聚体，结构如图 10-20 所示，两个十铌氧簇阴离子通过两个氧原子相连，形成 $[Nb_{20}O_{54}]^{8-}$。

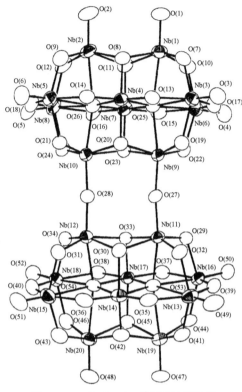

图 10-20 $[Nb_{20}O_{54}]^{8-}$ 原子连接图

10.3 十钼氧簇

10.3.1 合成

向七钼酸铵 $(NH_4)_6Mo_7O_{24}$ 浓水溶液中加入氨水，80℃结晶，几天后可得到组成为 $(NH_4)_8M_{10}O_{34}$ 的晶体[70]。

10.3.2 晶体结构

十钼酸铵晶体结构分析表明，十钼氧簇阴离子由 8 个 MoO_6 八面体和 2 个 MoO_4 四面体构成，结构如图 10-21 所示。其结构相当于八钼氧簇 γ 异构体利用阴离子中 2 个五配位四方锥构型钼原子的空位与两个 MoO_4 四面体相连，构成十钼氧簇阴离子。2 个 MoO_4 四面体通过共顶点和 $\gamma\text{-}Mo_8O_{26}^{4-}$ 相连。根据连接钼原子个数的差别，阴离子中的氧原子可分为：端基氧原子（O_t），共 20 个；二重桥氧原子（O_b），共 8 个；三重桥氧原子（O_c），共 4 个；四重桥氧原子（O_d），共 2 个。按照键合氧原子的种类和钼原子的构型，10 个钼原子可分为四类：第一类钼原子，连接 3 个端基氧原子和 1 个二重桥氧原子，为四面体构型，共 2 个；第二类钼原子，连接 2 个端基氧原子、2 个二重桥氧原子和 2 个三重桥氧原子，共 2 个；第三类连接 2 个端基氧原子、2 个二重桥氧原子、1 个三重桥氧原子和 1 个四重桥氧原子，共 4 个；第四类连接 1 个端基氧原子、1 个二重桥氧原子、2 个三重桥氧原子和 1 个四重桥氧原子，共 2 个。

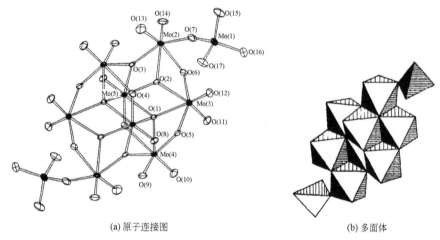

(a) 原子连接图　　　　　　　　　(b) 多面体

图 10-21　$[Mo_{10}O_{34}]^{8-}$ 原子连接图和多面体

10.3.3　相关化合物

目前报道的由十钼氧簇形成的化合物还很少，Yamase 等人报道了 $[NH_3Me]_8$ $[Mo_8O_{26}(MoO_4)_2]\cdot 2H_2O$ 的合成，并研究了光化学性质[71,72]。

在十钼氧簇阴离子中，由于阴离子构型的不同，其中氧原子的个数也会不同。在胡长文等人报道的化合物 $Na_4Mo_{10}O_{32}\cdot 8H_2O$ 中[73]，阴离子也含有 10 个钼原子，但阴离子结构却与前面描述的十钼氧簇阴离子差别很大，而与杂多阴离子的 Waugh 结构相似，结构如图 10-22 所示。阴离子中只有 32 个氧原子。

在化合物 $(NH_4)_2[Y(H_2O)_5]_2$ $[Mo_6^{VI}Mo_4^{V}O_{30}(CH_3COO)_4]\cdot 16H_2O$ 中[74]，由于稀土离子 Y^{3+} 和羧酸根的参与，阴离子表现为一种新的结构

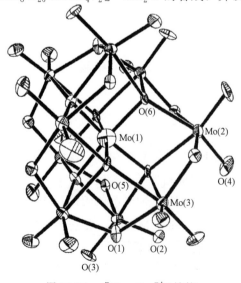

图 10-22　$[Mo_{10}O_{32}]^{4-}$ 结构

类型。图 10-23 为略去配位 Y^{3+} 及配位水后阴离子的结构。阴离子中共有 38 个氧原子参与了结构的形成,阴离子中钼原子的构型分为两种,8 个六配位的钼氧八面体通过共边和共顶点相连构成一个环,2 个 MoO_5 四方锥在环的内侧通过共边与 MoO_6 八面体构成的环相连,由于阴离子负电荷太高,阴离子通过端基氧原子与 Y^{3+} 结合,以增强阴离子的稳定性。在许林等人报道的化合物 [Na_8($Mo_{10}O_{32}EDTA$)(H_2O)$_{35}$]$_n$ 和 (NH_4)$_{8n}$[$Mo_{10}O_{32}PDTA$]$_n$(H_2O)$_{30n}$ 中[75],阴离子骨架结构与十钼氧簇阴离子 $Mo_8O_{34}^{8-}$ 相似,但与十钼氧簇阴离子不同的是,十钼氧簇阴离子中两个四面体构型的钼原子,在所报道的化合物中,结合了有机配体中的一个氮原子和一个氧原子,转变成了八面体构型(图 10-24)。

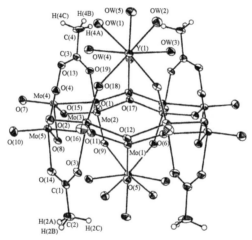

图 10-23　[$Y(H_2O)_5$]$_2$[$Mo_6^{VI}Mo_4^VO_{30}(CH_3COO)_4$]$^{2-}$ 结构

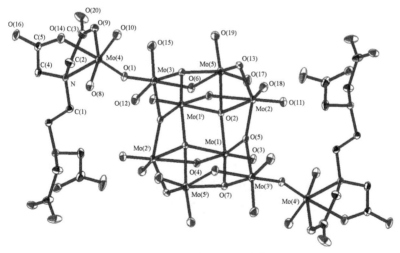

图 10-24　[$Mo_{10}O_{32}(EDTA)$]$_n^{8n-}$ 结构

以 $(n\text{-}Bu_4N)_2[Mo_5O_{13}(OMe)_4(NO)\{Na(MeOH)\}]$ 和 $(n\text{-}Bu_4N)_3[Mo_5O_{13}(OMe)_4(NO)\{Na(MeOH)\}]$ 为原料，可分别制得分子组成为 $(n\text{-}Bu_4N)[Mo_{10}O_{25}(OMe)_6(NO)]$ 和 $(n\text{-}Bu_4N)_2[Mo_{10}O_{24}(OMe)_7(NO)]$ 的化合物[76]。在这些化合物中，由于甲氧基和 NO 的配位，使得化合物阴离子结构与十钼氧簇 $Mo_{10}O_{34}^{8-}$ 结构相比差别更大（图 10-25）。

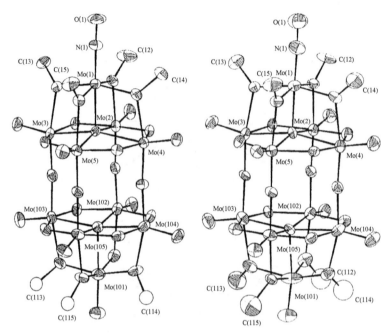

图 10-25 $[Mo_{10}O_{25}(OMe)_6(NO)]^-$ 和 $[Mo_{10}O_{24}(OMe)_7(NO)]^{2-}$ 结构

10.4 十钨氧簇

10.4.1 合成

1969 年，Chauveau 等人报道了 $K_4W_{10}O_{32}\cdot 4H_2O$ 的合成[77]，在后续的研究中，又有一些由十钨氧簇阴离子形成的无机盐和有机铵盐的报道[78-83]。

十钨氧簇四丁基铵盐的合成如下：16g$Na_2WO_4\cdot 2H_2O$ 溶于 100mL 水和 33.5mL 3mol/L 盐酸的混合溶液中，蒸沸几分钟后，向澄清的溶液中加入溴化四丁基铵水溶液（6.4g/10mL），收集沉淀，用沸水和乙醇洗涤，干燥，在热的二甲基甲酰胺中重结晶。得到黄色块状晶体$[(C_4H_9)_4N]_4W_{10}O_{32}$[78]。钠盐的制备方法与钾盐相似[77]。600mL 100g $Na_2WO_4\cdot 2H_2O$ 的水溶液与 600mL 1.0mol/L 盐酸的水溶液混合，回流 20s，加入过量固体 NaCl（300g），混合物再回流 20s，然后快速冷却，在冷冻（-10℃）下过夜，倾去清液，加入 300mL CH₃CN，混合物加热回流大约 5min，冷却后过滤，在蒸发皿中缓慢加热滤液直到第一个黄绿色晶体析出，可得到组成为 $Na_4W_{10}O_{32} \cdot xH_2O \cdot CH_3CN$ 的化合物[81]。

10.4.2 谱学表征

（1）红外光谱

十钨氧簇阴离子 $[W_{10}O_{32}]^{4-}$ 的红外光谱如图 10-26 所示，谱图上出现三个吸收峰，分别位于 960cm⁻¹、895cm⁻¹ 和 805cm⁻¹，其中位于 960cm⁻¹ 的峰为 W=O 双键的特征吸收峰，位于 895cm⁻¹ 和 805cm⁻¹ 的为 W—O—W 键的吸收峰。

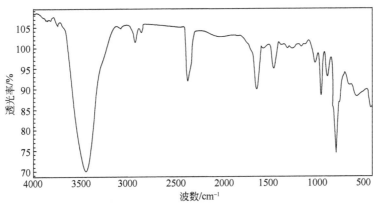

图 10-26　$[Ru(bpy)_3]_2[W_{10}O_{32}] \cdot 3DMSO$ 红外光谱

（2）电子光谱

图 10-27 为十钨氧簇阴离子 $[W_{10}O_{32}]^{4-}$ 在丙二醇-1,2-二碳酸酯中的电子光谱。图中在 265nm 和 325nm 附近出现两个吸收峰。265nm 附近的吸收峰对应于阴离子中桥氧的荷移跃迁吸收峰，这在大部分多金属氧簇阴离子的电子光谱图中均可观察到，但位于 325nm 附近的吸收峰却只在 Dawson 结构 $[P_2W_{18}O_{62}]^{6-}$ 和十钨氧簇阴离子中出现，这可能与这两种阴离子中存在近似性的 W—O—W 桥键相关[84]。理论计算结果也

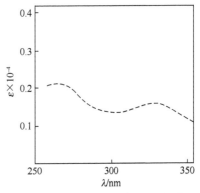

图 10-27　$[W_{10}O_{32}]^{4-}$ 电子光谱

表明，位于 325nm 附近的吸收峰和十钨氧簇阴离子中位于赤道位的钨原子相关[85]。

10.4.3 晶体结构

1973 年，Fuchs 报道了含有十钨氧簇阴离子 $[W_{10}O_{32}]^{4-}$ 的化合物 $[HN(C_4H_9)_3]_4W_{10}O_{32}$ 中阴离子的结构[82]，并在后续的工作中对化合物结构进行了研究[83]，其阴离子结构如图 10-28 所示。十钨氧簇阴离子可以看作是由两个单缺位的 Lindqvist 结构阴离子 $[W_5O_{18}]^{6-}$ 通过共顶点相连而成的。十个钨原子可分为两类，八个位于赤道位，两个位于极位。八个位于赤道位的钨原子分为两组，每组内钨原子间通过桥氧两两相连，构成一个钨氧八面体四元环，两个四元环通过四个共用氧原子相互连接，然后又通过位于两端的氧原子分别与位于中心的氧原子相连，构成柱状结构阴离子。每一个钨原子分别连接一个端氧、四个二重桥氧和一个五重桥氧原子。阴离子中的氧原子按照不同的配位环境可分为三类：①只与一个钨原子相连的端基氧原子，这类氧原子共有十个；②与两个钨原子相连的二重桥氧原子，这类氧原子共有二十个；③与五个钨原子相连的五重桥氧原子，这类氧原子共有两个。整个阴离子具有 D_{4h} 对称性。

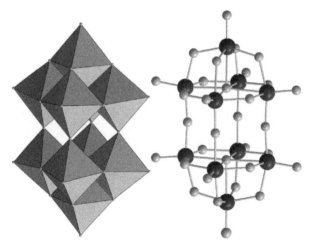

图 10-28　十钨氧簇阴离子 $[W_{10}O_{32}]^{4-}$ 结构

10.4.4 相关化合物

目前，由十钨氧簇阴离子形成的简单无机盐和有机铵盐有多篇报道[78-83,86,87]，但由十钨氧簇阴离子和配离子或其他类型的阳离子形成的化合物报道还不多。将 0.2g $[Ru(bpy)_3]Cl_2·6H_2O$ 溶于 20mL DMSO，搅拌下向其中滴加溶有 2g $(Bu_4N)_4W_{10}O_{32}$ 的 30mL 乙腈溶液，80℃下搅拌 1h，放置可得到分子组成为 $[Ru(bpy)_3]_2[W_{10}O_{32}]·3DMSO$ 的化合物[88]。化合物分子结构如图 10-29 所示，由一个十钨氧簇阴离子、两个 $[Ru(bpy)_3]_2^{2+}$ 配离子和三个

DMSO 溶剂分子组成，十钨氧簇阴离子和 $[Ru(bpy)_3]^{2+}$ 配离子之间通过静电引力相结合。将 $[C_2B_{10}H_{11}CH_2NH_3]Cl$ 与钨酸钠在酸性条件下的水溶液中反应，可得到分子组成为 $[C_2B_{10}H_{11}CH_2NHCH(CH_3)_2]_4[W_{10}O_{32}][H_2O]_2[(CH_3)_2CO]_4$ 的白色沉淀[89]，将该沉淀在丙酮/环己烷混合溶剂中重结晶，可得到适用于单晶分析的晶体，化合物分子结构如图 10-30 所示。在化合物中，阴、阳离子之间通过静电引力相结合。

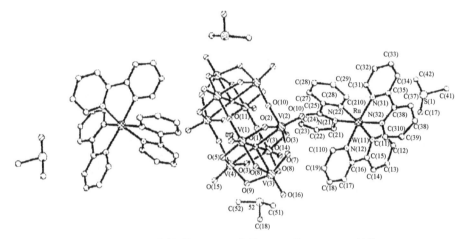

图 10-29　化合物 $[Ru(bpy)_3]_2[W_{10}O_{32}] \cdot 3DMSO$ 结构

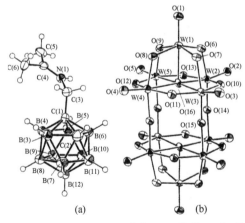

图 10-30　化合物 $[C_2B_{10}H_{11}CH_2NHCH(CH_3)_2]_4[W_{10}O_{32}][H_2O]_2[(CH_3)_2CO]_4$ 分子结构

二苯并 18-冠-6 的乙腈溶液与 $[Bu_4N]_4W_{10}O_{32}$ 的甲醇溶液反应，用二甲基甲酰胺重结晶后，可得到分子组成为 $[(DB_{18}C_6)(DMF)_2Na]_4 \cdot W_{10}O_{32} \cdot 2DMF \cdot 2H_2O$ 的化合物[90]。在化合物中，十钨氧簇阴离子和 $[(DB_{18}C_6)(DMF)_2]^+$ 通过静电引力相结合。将 $Cu(NO_3)_2 \cdot 3H_2O$、2,2'-联吡啶、$Na_2WO_4 \cdot 2H_2O$ 和 H_2O 按 1∶1∶3∶1900 的摩尔比混合，用 HNO_3 调 pH=5，在 140℃下反应 4d，

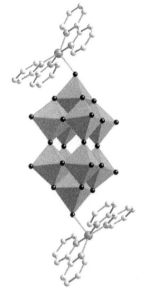

得到分子组成为 $[\{Cu(bpy)_2\}_2W_{10}O_{32}]\cdot 2H_2O$ 的化合物[91]，在该化合物中，十钨氧簇阴离子通过位于极位的钨原子的端基氧与 $[Cu(bpy)_2]^{2+}$ 配离子配位，形成担载结构的化合物（图 10-31）。

$[Bu_4N]_4W_{10}O_{32}$ 与 $Ce_2(C_2O_4)_3$ 反应，可得到结构单元为 $[Ce(H_2O)(DMF)_6(W_{10}O_{32})]\cdot(DMF)\cdot(CH_3CH_2OH)$ 的化合物[92]。值得注意的是，该化合物在生成的过程中 Ce^{3+} 转变为 Ce^{4+}。在化合物中，十钨氧簇阴离子作为二齿配体，通过位于极位和赤道位的两个钨原子的端基氧与两个 $[Ce(H_2O)(DMF)_6]^{4+}$ 相连，而每一个 $[Ce(H_2O)(DMF)_6]^{4+}$ 配离子又以两个空位分别与两个十钨氧簇阴离子相连，构成一维螺旋链状结构（图 10-32）。

图 10-31 化合物 $[\{Cu(bpy)_2\}_2W_{10}O_{32}]\cdot 2H_2O$ 结构

用 $[(en^*)Pd(4,4'\text{-}bpy)]_3(NO_3)_6$ 和 $[(en^*)Pd(4,4'\text{-}bpy)]_4(NO_3)_8$（$en^*$ 为 N,N',N',N'-四甲基

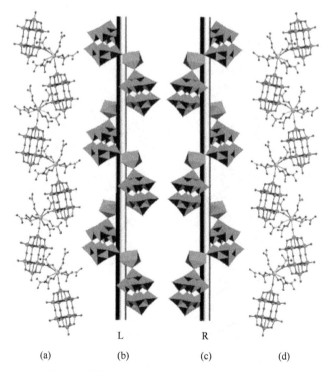

图 10-32 化合物 $[Ce(H_2O)(DMF)_6(W_{10}O_{32})]\cdot(DMF)\cdot(CH_3CH_2OH)$ 构成的一维螺旋链状结构

乙二胺）分别与 $[Bu_4N]_4[W_{10}O_{32}]$ 反应，可分别得到组成为 $\{[(en^*)Pd(4,4'-bpy)]_4[\supset W_{10}O_{32}]\}[W_{10}O_{32}]$ 和 $\{[(en^*)Pd(4,4'-bpy)]_4[W_{10}O_{32}]_2$ 的化合物[93]。化合物结构如图 10-33 和图 10-34 所示。在化合物中，阴阳离子之间通过静电引力相结合。有意思的是，在$[(en^*)Pd(4,4'-bpy)]_4(NO_3)_8$ 参与的反应中，十钨氧簇阴离子进入了配阳离子的四方框架内。

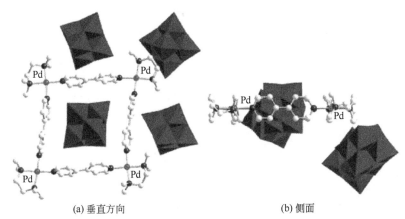

(a) 垂直方向　　　　　(b) 侧面

图 10-33　化合物 $\{[(en^*)Pd(4,4'-bpy)]_4[\supset W_{10}O_{32}]\}[W_{10}O_{32}]$ 结构

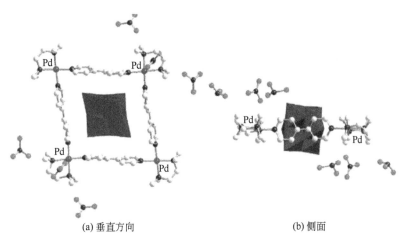

(a) 垂直方向　　　　　(b) 侧面

图 10-34　化合物 $[(en^*)Pd(4,4'-bpy)]_4[W_{10}O_{32}]_2$ 结构

$Na_2WO_4 \cdot 2H_2O$、$Cu(NO_3)_2 \cdot 6H_2O$、2,2'-联吡啶和 DL-羟基琥珀酸在水热条件下反应，可得到分子组成为 $[Cu(2,2'-bpy)_2]_2[Na_2W_6^VW_4^{VI}O_{30}] \cdot 2H_2O$ 的化合物[94]，在该化合物中，阴离子与配阳离子通过配位键相连（图 10-35），其阴离子与十钨氧簇阴离子结构十分相似，只是两个 Na^+ 取代了十钨氧簇阴离子中位于笼内的两个氧原子。

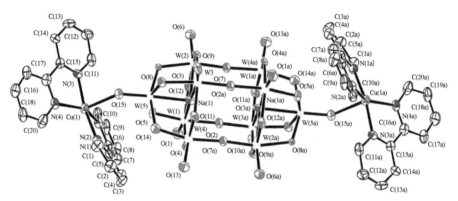

图 10-35 化合物 $[Cu(2,2'-bpy)_2]_2[Na_2W_6^V W_4^{VI} O_{30}] \cdot 2H_2O$ 结构

参考文献

[1] Radau C. Salts of vanadic acid. Chem Zent, 1888: 1378.

[2] Rossotti F J C, Rossotti H. Equilibrium studies of polyanions I. Isopolyvanadates in acidic media. Acta Chem Scand, 1956, 10 (6): 957-984.

[3] Naumann A W, Hallada C J. A study of the chemistry of the polyvanadates using salt cryoscopy. Inorg Chem, 1964, 3 (1): 70-77.

[4] Evans H T Jr. The molecular structure of the isopoly complex ion, decavanadate ($V_{10}O_{28}^{6-}$). Inorg Chem, 1966, 5 (6): 967-977.

[5] Clare B W, Kepert D L, Watts D W. Kinetic study of the acid decomposition of decavanadate. J C S Dalton Trans, 1973: 2479-2480.

[6] Wery A S J, Gutierrez-Zorrila J M, Luque A, Roman P. Influence of protonation on crystal packing and thermal behaviour of tert-butylammonium decavanadates. Polyhedron, 1996, 15 (24): 4555-4564.

[7] Rigotti G, Punte G, Rivero B E, Escobar M E, Baran E I. Crystal data and vibrational spectra of the rare earth decavanadates. J Inorg Nucl Chem, 1981, 43: 2811.

[8] Žúrková L, Vávra R. Synthesis and properties of propylammonium polyvanadates. Monatsh Chem, 1993, 124: 619.

[9] Žúrková L, Tatiersky J, Fejdi P. Synthesis and properties of dipropylammonium and 1,3-propanediammonium polyvanadates. Monatsh Chem, 1998, 129: 577-584.

[10] Evans H T Jr, Swallow A G, Barnes W H. The structure of the decavanadate ion. J Am Chem Soc, 1964, 86: 4209-4210.

[11] Klemperer W G, Shum W. Charge distribution in large polyoxoanions: determination of protonation sites in $V_{10}O_{28}^{6-}$ by ^{17}O nuclear magnetic resonance. J Am Chem Soc, 1977, 99 (10): 3544-3545.

[12] Day V W, Klemperer W G, Maltbie D J. Where are the protons in $H_3V_{10}O_{28}^{3-}$?. J Am

Chem Soc, 1987, 109 (10): 2991-3002.

[13] Kempf J Y, Rohmer M M, Poblet J M, Bo C, Bénard M. Relative basicities of the oxygen sites in $[V_{10}O_{28}]^{6-}$. An analysis of the ab initio determined distributions of the electrostatic potential and of the laplacian of charge density. J Am Chem Soc, 1992, 114: 1136-1146.

[14] Román P, Aranzabe A, Luque A, Gutiérrez-Zorrilla J M, Martínez-Ripoll M. Effects of protonation in decavanadates: cryastal structure of tetrakis (n-hexylammonium) dihydrogendecavanadate (Ⅴ). J Chem Soc Dalton Trans, 1995: 2225-2231.

[15] Capparelli M V, Goodgame D M L, Hayman P B, Skapski A C. Protonation sites in the decavanadatelon: X-ray crystal structure of tetrakis-adenosinium dihydrodecavanadate (Ⅴ) undecahydrate. J Chem Soc Chem Commun, 1986: 776-777.

[16] Shao M C, Wang L F, Zhang Z Y, Tang Y Q. The crystal structure of the addition compound of decavanadic acid and 5,7-dimethyl-2,3-dihydro-1,4-diazepinium. Scientia Sinica Chimica (Series B), 1984: 137-148.

[17] Pecquenard B, Zavalij P Y, Whittingham M S. Tetrakis (tetramethylammonium) dihydro-gendecavanadate acetic acid 2.8-hydrate, $[N(CH_3)_4]_4[H_2V_{10}O_{28}] \cdot CH_3COOH \cdot 2.8H_2O$. Acta Cryst, 1988, C54: 1833-1835.

[18] Wang W, Zeng F L, Wang X, Tan M Y. Preparation and the crystal structure of a new tetra-N-butylammonium decavanadate. Polyhedron, 1996, 15 (2): 265-268.

[19] Farahbakhsh M, Schmidt H, Rehder D A. Decavandate sandwiched by diprotonated cryptands-222: model for the vanadate-ionophore interaction. Chem Ber/Recueil, 1997, 130: 1123-1127.

[20] Hu N H, Tokuno T, Aoki K. Novel structural aspects of oxythiamine, an antagonist of thiamine. Crystal structures of three salts, (oxythiamineH)(picrolonate)$_2 \cdot 2H_2O$, (oxythiamineH)(PtCl$_6$), (oxythiamineH)$_2$(V$_{10}$O$_{28}$H$_2$) $\cdot 8H_2O$, and a metal complex Pt(oxythiamine)Cl$_3 \cdot H_2O$. Inorg Chim Acta, 1999, 295: 71-83.

[21] Kondo M, Fujimoto K, Okubo T, Asami A, Noro S, Kitagawa S, Ishii T, Matsuzaka H. Novel extended linear structure of decavanadate anions linked by bis (4-pyridinium) disulfide (H$_2$dpds), $\{(H_2dpds)_2[V_{10}O_{26}(OH)_2] \cdot 10H_2O\}_n$. Chem Lett, 1999: 291-292.

[22] Biagioli M, Strinna-Erre L, Micera G, Panzanelli A, Zema M. Tetrahydrogendecavanadate (Ⅴ) and its binding to glycylglycine. Inorg Chem Commun, 1999, 2: 214-217.

[23] Chinea E, Dakternieks D, Duthie A, Ghilardi C A, Gili P, Mederos A, Midollini S, Orlandini A. Synthesis and characterization of $(NH_4)_4[H_4V_{10}O_{28}][NTPH_2]_2 \cdot 4H_2O$ (NTPH$_3$=N(CH$_2$CH$_2$COOH)$_3$). Inorg Chim Acta, 2000, 298: 172-177.

[24] Rakovský E, Žúrková L, Marek J. Synthesis, crystal structure, and IR spectroscopic characterization of 1,6-hexanediammonium dihydrogendecavanadate. Monatsh Chem, 2002, 133: 277-283.

[25] Correia I, Avecilla F, Marcão S, Pessoa J C. Structural studies of decavanadate compounds with organic molecules and inorganic ions in their crystal packing. Inorg Chim Acta, 2004, 357: 4476-4487.

[26] Kumagai H, Arishima M, Kitagawa S, Ymada K, Kawata S, Kaizaki S. New hydrogen bond-supported 3-D molecular assembly from polyoxovanadate and tetramethylbiimidazole. Inorg Chem, 2002, 41: 1989-1992.

[27] Zhai H J, Liu S X, Peng J, Hu N H, Jia H Q. Synthesis, crystal structure, and thermal property of a novel supramolecular assembly: $(NH_4)_2(C_8H_{10}N_4O_2)_4[H_4V_{10}O_{28}] \cdot 2H_2O$, constructed from decavanadate and caffeine. J Chem Cryst, 2004, 34 (8): 541-548.

[28] Pavani K, Upreti S, Ramanan A. Two new polyoxovanadate clusters templated through cysteamine. J Chem Sci, 2006, 118 (2): 159-164.

[29] Liu J T, Wang X H, Liu J F. Synthesis and crystal structure of a novel compound of $(C_2N_2H_8O_2)H_6V_{10}O_{28}$. Chinese Chem Lett, 2004, 15 (7): 859-862.

[30] Wang D R, Zhang W J, Grüning K, Rehder D. Inorganic/organic hybrid salts derived from polyoxovanadates and macrocyclic (O_xN_y) cations. J Mol Struct, 2003, 656: 79-91.

[31] Bukietyńska K, Krot K, Starynowicz P. Distortion of the decavanadate polyhedron and the role of hydrogen bonding in dimethylammonium decavanadate. Trans Met Chem, 2001, 26: 311-314.

[32] Wang X, Liu H X, Xu X X, You X Z. The crystal structure of the addition compound of decavanadate and guanidine. Polyhedron, 1993, 12 (1): 77-81.

[33] Crans D C, Mahroof-Tahir M, Anderson O P, Miller M M. X-ray structure of $(NH_4)_6(Gly-Gly)_2V_{10}O_{28} \cdot 4H_2O$: model studies for polyoxometalate-protein interaction. Inorg Chem, 1994, 33: 5586-5590.

[34] Shan Y K, Huang S D. Hydrothermal crystallization and X-ray structure determination of a new decavanadate with mixed cations $[Mn(H_2O)_6]_2[N(CH_3)_4]_2[V_{10}O_{28}] \cdot 2H_2O$. J Chem Cryst, 1999, 29 (1): 93-97.

[35] Naruke H, Yamase T, Kaneko M. X-ray structural characterization of $[Eu(H_2O)_8]_2[V_{10}O_{28}] \cdot 8H_2O$. Bull Chem Soc Jpn, 1999, 72: 1775-1779.

[36] Peng X H, Li Y Z, Cai L X, Wang L F, Wu J G. A novel polyoxometalatesupramolecular compound: $[La(H_2O)_9]_2[V_{10}O_{28}] \cdot 8H_2O$. Acta Cryst, 2002, E58: i111-i113.

[37] Zhang X M, Chen X M. Three-dimensional supramolecular arrays supported by decavanadate clusters: syntheses and crystal structures of $(NH_4)_2[M(dod)(H_2O)_4]_2V_{10}O_{28} \cdot 6H_2O$, $(M=Zn, Mn)$. Inorg Chem Commun, 2003, 6: 206-209.

[38] Ninclaus C, Riou D, Férey G. A new decavanadate dihydrate templated by ethylenediamine. Acta Cryst, 1996, C52: 512-514.

[39] Zavalij P Y, Chirayil T, Whittingham M S, Pecharsky V K, Jacobson R A. A new decavanadate with mixed cations, $[Li(H_2O)_4]_2[N(CH_3)_4]_4[V_{10}O_{28}] \cdot 4H_2O$. Acta Cryst, 1997, C53: 170-171.

[40] Strukan N, Cindrić M, Kamenar B. Ca$_2$(H$_3$O)$_2$[V$_{10}$O$_{28}$]·16H$_2$O. Acta Cryst, 1999, C55: 291-293.

[41] Fratzky D, Schneider M, Rabe S, Meisel M. (NH$_4$)$_4$Na$_2$[V$_{10}$O$_{28}$]·10H$_2$O. Acta Cryst, 2000, C56: 740-741.

[42] Piro O E, Varetti E L, Brandán S A, Altabef A B. A crystallographic and vibrational study of cesium di-μ-aqua bis [tetraaquasodium(I)] decavanadate, Cs$_4$[Na$_2$(H$_2$O)$_{10}$](V$_{10}$O$_{28}$). J Chem Cryst, 2003, 33 (1): 57-63.

[43] 詹晓平, 卢灿忠, 杨文斌, 马宏伟, 吴传德, 张全争, 蒋晓瑜. 两个具有相似{V$_{10}$}阴离子骨架的化合物的合成和结构表征. 无机化学学报, 2003, 19 (8): 831-836.

[44] Ksiksi R, Graia M, Driss A et Jouini T. Décavanadatesel double de dilithium et tétra-ammonium décahydrate, (NH$_4$)$_4$Li$_2$[V$_{10}$O$_{28}$]·10H$_2$O. Acta Cryst, 2004, E60: i105-i107.

[45] Chen L, Lin Z Z, Jiang F L, Yuan D Q, Hong M C. Synthesis and crystal structure of a 3-D hydrogen-bonded supramolecular decavanadate compound [Habo]$_6$[V$_{10}$O$_{28}$]·2C$_3$H$_7$OH·2H$_2$O. Chin J Struct Chem, 2005, 24 (10): 1186-1192.

[46] Zhu C Y, Li Y T, Wu Z Y, Jiang M. Tris (propane-1, 3-diammonium) decavanadate 5.5-hydrate. Acta Cryst, 2006, E62: m3092-m3095.

[47] Rakovský E, Gyepes R. Butane-1, 4-diammonium decavanadate (Ⅴ) hexahydrate. Acta Cryst, 2006, E62: m2108-m2110.

[48] Zhao Q H, Du L, Fang R B. Bis (trimethylenediammonium) hexaaquacopper (Ⅱ) decavanadateheptahydrate. Acta Cryst, 2006, E62: m360-m362.

[49] Xie A L, Ma C A, Wang L B. Diammoniumbis [hexaaquanickel (Ⅱ)] decavanadatetetrahydrate, (NH$_4$)$_2$[Ni(H$_2$O)$_6$]$_2$V$_{10}$O$_{28}$·4H$_2$O. Acta Cryst, 2006, E62: i1-i3.

[50] Zhu C Y, Li Y T, Wu Z Y, Xiao N Y. Tris (N,N-dimethylethane-1, 2-diammonium) decavanadate (Ⅴ) pentahydrate. Acta Cryst, 2007, E63: m547-m549.

[51] Zhu C Y, Li Y T, Wu J D, Wu Z Y, Wu X. Tris [(3-aminopropyl) dimethyl-ammonium] decavanadate (Ⅴ) hexa-hydrate. Acta Cryst, 2007, E63: m1777-m1778.

[52] Wang L, Sun X P, Liu M L, Gao Y Q, Gu W, Liu X. Syntheses, structures and properties of three heteonuclear complexes containing [V$_{10}$O$_{28}$]$^{6-}$ units. J Clust Sci, 2008, 19 (3): 531-542.

[53] Khan M I, Tabussum S, Zheng C. Mixed-metal oxide phases containing decavanadate clusters: synthesis and crystal structures of {(H$_2$O)$_2$K-μ-(H$_2$O)$_3$-M(H$_2$O)$_3$}$_2$[V$_{10}$O$_{28}$](M=Co, Ni). J Clust Sci, 2001, 12 (4): 583-594.

[54] Higami T, Hashimoto M, Okeya S. [Ni(H$_2$O)$_6$]$_2$[Na(H$_2$O)$_3$]$_2$[V$_{10}$O$_{28}$]·4H$_2$O, bis (nickel hexahydrate) bis (sodium trihydrate) decavanadate tetrahydrate. Acta Cryst, 2002, C58: i144-i146.

[55] Lida A, Ozeki T. Mg$_2$Na$_2$V$_{10}$O$_{28}$·20H$_2$O and Mg$_3$V$_{10}$O$_{28}$·28H$_2$O. Acta Cryst, 2004, C60: i43-i46.

[56] Zhang X M, Wu H S, Chen X M. A three-dimensional organic-inorganic hybrid material

supported by decavanadate clusters and Na-O chains: synthesis and crystal structure of $[Na_6(H_2O)_{16}(dod)_4V_{10}O_{28}]$. Chin J Struct Chem, 2004, 23 (4): 407-412.

[57] Ma C A, Xie A L, Wang L B. Polymeric bis (lithium disodium nonahydrate) decavanadate, $\{[LiNa_2(H_2O)_9]_2[V_{10}O_{28}]\}_n$. Acta Cryst, 2005, E61: i185-i187.

[58] Lida A, Ozeki T. $Cu_3V_{10}O_{28} \cdot 24H_2O$ and $CuNa_4V_{10}O_{28} \cdot 23H_2O$. Acta Cryst, 2003, C59: i41-i44.

[59] Miras H N, Raptis R G, Lalioti N, Sigalas M P, Baran P, Kabanos T A. A novel series of vanadium-sulfite polyoxometalates: synthesis, structural, and physical studies. Chem Eur J, 2005, 11: 2295-2306.

[60] Graia M, Ksiksi R, Driss A. Preparation and characterization of $[Zn(H_2O)_6][Zn_2V_{10}O_{28}(H_2O)_{10}] \cdot 6H_2O$ single crystals with a novel metallic heteropolyoxoanion. J Chem Crystallogr, 2008, 38 (11): 855-859.

[61] Li T, Lü J, Gao S, Li F, Cao R. Inorganic-organic hybrid with 3D surpramolecular channel assembled through C-H⋯O interactions based on the decavanadate. Chem Lett, 2007, 36 (3): 356-357.

[62] Khan M I, Chen Q, Goshorn D P, Hope H, Parkin S, Zubieta J. Polyoxo alkoxides of vanadium: the structures of the decanuclear vanadium(IV) clusters $[V_{10}O_{16}\{CH_3CH_2C(CH_2O)_3\}_4]^{4-}$ and $[V_{10}O_{13}\{CH_3CH_2C(CH_2O)_3\}_5]^{-}$. J Am Chem Soc, 1992, 114: 3341-3346.

[63] Khan M I, Chen Q, Goshorn D P, Zubieta J. Polyoxo alkoxide clusters of vanadium: structural characterization of the decavanadate core in the "fully reduced" vanadium(IV) species $[V_{10}O_{16}\{(OCH_2)_3CCH_2CH_3\}_4]^{4-}$ and $[V_{10}O_{14}(OH)_2\{(OCH_2)_3CCH_2OH\}_4]^{2-}$ and in the mixed-valence clusters $[V_8^{IV}V_2^VO_{16}\{(OCH_2)_3CR\}_4]^{2-}$ (R=—CH_2CH_3, —CH_3). Inorg Chem, 1993, 32: 672-680.

[64] Khan M I, Chen Q, Zubieta J. Synthesis and crystal and molecular structure of $(NH_4)_4[V_{10}O_{16}\{EtC(CH_2O)_3\}_4] \cdot 4H_2O$, a decavanadyl cluster. J Chem Soc Chem Commun, 1992: 305.

[65] 李阳光, 栾国有, 王恩波, 王树涛, 胡长文, 胡宁海, 贾恒庆. 新型笼状同多钒酸盐 $[Ni(1,10'-phen)_3]_2 \cdot [V_2^{IV}V_8^VO_{26}]$ 的水热合成和晶体结构. 高等学校化学学报, 2003, 24 (1): 34-36.

[66] Graeber E J, Morosin B. The molecular configuration of the decaniobate ion ($Nb_{10}O_{28}^{6-}$). Acta Cryst, 1977, B33: 2137-2143.

[67] Shen L, Li C H, Chi Y N, Hu C W. Zn $(2,2'-bipy)_2$/Co $(2,2'-bipy)_2$ linked decaniobate $[Nb_{10}O_{28}]^{6-}$ clusters-zigzag neutral chains. Inorg Chem Commun, 2008, 11: 992-994.

[68] Ohlin C A, Villa E M, Casey W H. One-pot synthesis of the decaniobate salt $[N(CH_3)_4]_6[Nb_{10}O_{28}] \cdot 6H_2O$ from hydrous niobium oxide. Inorg Chim Acta, 2009, 362: 1391-1392.

[69] Maekawa M, Ozawa Y, Yagasaki A. Icosaniobate: a new member of the isoniobate

family. Inorg Chem, 2006, 45: 9608-9609.

[70] Fuchs J, Hartl H, Hunnius W D, Mahjour S. Anion structure of ammonium decamolybdate $(NH_4)_8Mo_{10}O_{34}$. Angew Chem Internat Edit, 1975, 14 (9): 644.

[71] Bharadwaj P K, Ohashi Y, Sasada Y, Yamase T. Structure of octakis (methylammonium) decamolybdate (8-) dihydrate. Acta Crystallogr, 1986, C42: 545-547.

[72] Yamase T. Photochemical studies of the alkylammonium molybdates. Part 7. Octahedral sites for multi-electron reduction of $[Mo_8O_{26}(MoO_4)_2]^{8-}$. J Chem Soc Dalton Trans, 1985: 2585-2590.

[73] Feng L, Wang Y, Qi Y, Hu C, Xu Y, Wang E. Synthesis and crystal structure of the first Waugh-type isopolyoxomolybdate $Na_4Mo_{10}O_{32} \cdot 8H_2O$. J Mol Struct, 2003, 645: 231-234.

[74] Liu G, Zhang S W, Tang Y Q. Synthesis and characterization of a novel ring-like decamolybdate. J Chem Soc Dalton Trans, 2002: 2036-2039.

[75] Gao G G, Xu L, Qu X S, Liu H, Yang Y Y. New approach to the synthesis of an organopolymolybdate polymer in aqueous media by linkage of multicarboxylic ligands. Inorg Chem, 2008, 47: 3402-3407.

[76] Proust A, Robert F, Gouzerh P, Chen Q, Zubieta J. Reduced nitrosyl polyoxomolybdates with the hitherto unknown decamolybdate Y structure: preparation and crystal and electronic structures of the two-electron reduced $[Mo_{10}O_{25}(OMe)_6(NO)]^-$ and the four-electron reduced $[Mo_{10}O_{24}(OMe)_7(NO)]^{2-}$. J Am Chem Soc, 1997, 119: 3523-3535.

[77] Chauveau F, Boyer M, LeMeur B. Preparations and properties of two isopolystungstic acids. Comptes Rendus des Seances de l'Academie des Sciences: Serie C, 1969, 268: 479-482.

[78] Chemseddine A, Sanchez C, Livage J, Launay J P, Fournier M. Electrichemical and photochemical reduction of decatungstate: a reinvestigation. Inorg Chem, 1984, 23: 2609-2613.

[79] Filowitz M, Ho R K C, Klemperer W G, Shum W. Oxygen-17 nuclear magnetic resonance spectroscopy of polyoxometalates. 1. Sensitivity and resolution. Inorg Chem, 1979, 18 (1): 93-103.

[80] Nomiya K, Matsubara S, Yamashita K, Miwa M. Effect of countercations in catalytic photooxidation of isopropyl alcohol by decatungstate isopolyanion. Polyhedron, 1987, 6 (10): 1919-1921.

[81] Renneke R F, Pasquali M, Hill C L. Polyixometalate systems for the catalytic selective production of nonthermodynamic alkenes from alkanes. Nature of excited-state deactivation processes and control of subsequent thermal processes in polyocometalate photoredox chemistry. J Am Chem Soc, 1990, 112 (18): 6585-6594.

[82] Fuchs J, Hartl H, Schiller W. Anion structure of tributylammonium decatungstate $[HN(C_4H_9)_3]_4W_{10}O_{32}$. Angew Chem Internat Edit, 1973, 12 (5): 420.

[83] Fuchs J, Hartl H, Schiller W, Gerlach U. Die kristallstruktur des tributylammonium dekawolframats [$(C_4H_9)_3NH]_4W_{10}O_{32}$. Acta Cryst, 1976, B32: 740-749.

[84] Termes S C, Pope M T. Reduction of the decatungstate anion in nonaqueous solution and its confirmation as "polytungstate-Y". Inorg Chem, 1978, 17 (2): 500-501.

[85] Inoue M, Yamase T, Kazansky L P. NMR and UV spectra of lanthanide decatungstates $LnW_{10}O_{36}^{n-}$ and $W_{10}O_{32}^{4-}$: A study of some peculiarities in spectra by the extended Hückel MO method. Polyhedron, 2003, 22: 1183-1189.

[86] Nomiya K, Miyazaki T, Maeda K. Catalytic photooxidation of some secondary alcohols by decatungstate isopolyanion and Keggin-type dodecatungstophosphate heteropolyanion in homogeneous system under excess of oxygen and the effect of counterions on redox cycle of polyanions. Inorg Chim Acta, 1987, 127: 65-69.

[87] Nomiya K, Sugie Y, Miyazaki T, Miwa M. Catalysis by heteropolyscids-ix. Photocatalytic oxidation of isopropyl alcohol to acetone under oxygen using tetrabutylammonium decatungstate. Polyhedron, 1986, 5: 1267-1271.

[88] Han Z, Wang E, Luan G, Li Y, Hu C, Wang P, Hu N, Jia H. Synthesis and crystal structure of a novel compound constructed from tris- (2,2'-bipy) ruthenium(II) and decatungstate. Inorg Chem Commun, 2001, 4: 427-429.

[89] Macias R, Kennedy J D, Thornton-Pett M, Románb P. The "globule-globule" hybrid dicarbaborane -polyoxometallate salt, $[C_2B_{10}H_{11}CH_2NHCH(CH_3)_2]_4[W_{10}O_{32}][H_2O]_2[(CH_3)_2CO]_4$. Cryst Eng Comm, 2003, 5 (19): 93-95.

[90] Li Y, Wang E, Lu Y, Hu C, Hu N, Jia H. Synthesis, characterization and crystal structures of dibenzo-18-crown-6 sodium isopolytungstates. J Mol Struct, 2002, 607: 133-141.

[91] Devi R N, Burkholder E, Zubieta J. Hydrothermal synthesis of polyoxotungstate clusters, surface-modified with M(II) -organonitrogen subunits. Inorg Chim Acta, 2003, 348: 150-156.

[92] Liu C, Luo F, Liu N, Cui Y, Wang X, Wang E, Chen J. One-dimensional helical chain based on decatungstate and cerium organic-inorganic hybrid material. Crystal Growth Des, 2006, 6 (12): 2658-2660.

[93] Uehara K, Kasai K, Mizuno N. Syntheses and characterizations of palladium-based moecular triangle/square compounds and hybrid composites with polyoxometalates. Inorg Chem, 2007, 46: 2563-2570.

[94] Li T, Lu J, Gao S, Cao R. Hydrothermal synthesis, crystal structure and magnetic property of a novel mixed valence decatungstate: $[Cu(2,2'-bipy)_2]_2[Na_2W_6^VW_4^{VI}O_{30}]\cdot 2H_2O$. Chin J Struct Chem, 2008, 27 (1): 57-62.

第11章

高聚金属氧簇

由于多于十个金属核的聚金属氧簇阴离子结构类型比较零散，因此，把它们均归在本章讨论。从现有文献看，形成高聚金属氧簇的元素有钒、钼、钨、铌，铌仅有个例报道，而钽至今未见有高聚氧簇。

11.1 钒氧簇

11.1.1 十二钒氧簇

将 $[Bu_4N]_4[H_2V_{10}O_{28}]$ 的乙腈溶液回流 1～2min，向其中加入足量的乙醚，可析出沉淀，将沉淀用乙腈、乙酸乙酯体积比为 1∶2 的溶剂在 −5℃ 下重结晶，可得到分子组成为 $[Bu_4N]_4[CH_3CN\subset(V_{12}O_{32})]$ 的化合物[1]。适合用于 X 射线单晶分析的晶体可转化为四苯基𬭸盐得到，组成为 $[(C_6H_5)_4P]_4[V_{12}O_{32}]\cdot 4CH_3CN\cdot 4H_2O$。晶体结构分析表明，十二钒氧簇阴离子 $V_{12}O_{32}^{4-}$ 具有篮状的笼形结构，乙腈分子虚悬在笼的中间（图 11-1），阴离子中的钒原子均为五配位四方锥构型，阴离子中的 32 个氧原子可分为三类，有 12 个端基氧原子、12 个二重桥氧原子和 8 个三重桥氧原子，阴离子具有 C_{4v} 对称性。

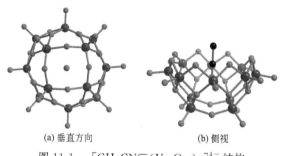

(a) 垂直方向　　　　(b) 侧视

图 11-1　$[CH_3CN \subset (V_{12}O_{32})]^{4-}$ 结构

偏钒酸钠溶液与水合肼 N_2H_5OH 反应，用氢氟酸调节酸度，可得到分子组成为 $Na_6[H_6V_{12}O_{30}F_2] \cdot 22H_2O$ 的化合物[2]，晶体结构分析表明，在该化合物中，十二钒氧簇阴离子结构与前述的篮状结构不同，是具有封闭的笼型结构 [图 11-2(a)]。在该阴离子中，十个钒原子被还原为 +4 价，剩余两个 +5 价。6 个 +4 价的钒分为两组分别与两个氟离子配位，构成六配位八面体构型，这 6 个 +4 价钒之间的桥氧原子均连接一个质子，另外 4 个 +4 价的钒和 2 个 +5 价的钒均为五配位四方锥构型，整个阴离子可以看作 6 个五配位的钒氧四方锥构成一个六元环，两组共边的六配位钒氧八面体分别位于环的两侧 [图 11-2(b)]。

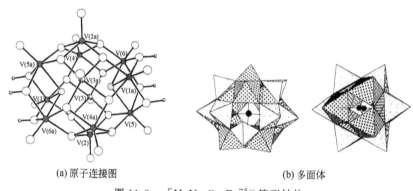

(a) 原子连接图　　　　(b) 多面体

图 11-2　$[H_6V_{12}O_{30}F_2]^{6-}$ 笼型结构

用 Et_4NVO_3 与一氧化氮（NO）在硝基甲烷中反应，可得到深棕色微晶，元素分析和单晶测定结果表明化合物分子组成为 $[Et_4N]_5[(NO)V_{12}O_{32}]$[3]，结构分析表明，该阴离子与化合物 $[Bu_4N]_4[CH_3CN \subset (V_{12}O_{32})]$ 中阴离子同构，只是在阴离子中间虚悬的 CH_3CN 分子换成了一个虚悬的硝酰阴离子 NO^-。

11.1.2　十三钒氧簇

现已报道的十三钒氧簇有两种结构类型。

将 $(Bu_4N)_4[H_2V_{10}O_{28}]$ 在乙腈中回流 1～2min，可得到十二钒氧簇阴离

子 $[CH_3CN(V_{12}O_{32})]^{4-}$ 构成的化合物 $[Bu_4N]_4[CH_3CN\subset(V_{12}O_{32})]$[1]，如果将 $(Bu_4N)_4[H_2V_{10}O_{28}]$ 乙腈溶液在干燥的氮气氛下回流 7h，则可得到十三钒氧簇阴离子 $[V_{13}O_{34}]^{3-}$ 的四丁基铵盐 $[Bu_4N]_3[V_{13}O_{34}]$[4]。其阴离子结构如图 11-3 所示，阴离子为层状结构，阴离子中所有钒原子均为六配位八面体构型，八面体之间采用紧密堆积方式。按照所处位置的不同，阴离子中的钒原子可分为四类：①位于中心位置的钒原子，只有 1 个，连接三个六重桥氧和三个三重桥氧；②位于次中心位置的钒原子，这类钒原子共有 3 个，每个钒原子连接两个六重桥氧、两个三重桥氧和两个二重桥氧；③位于次外层的钒原子，这类钒原子共有 6 个，每个钒原子连接一个六重桥氧、两个三重桥氧、两个二重桥氧和一个端基氧原子；④位于最外层的钒原子，这类钒原子共有 3 个，每个钒原子连接一个六重桥氧、四个二重桥氧和一个端基氧原子。阴离子中的氧原子按照配位环境的不同可分为四类：①只与一个钒原子相连的氧原子，共 9 个；②与两个钒原子相连的二重桥氧原子，共 15 个；③与三个钒原子相连的三重桥氧原子，共 7 个；④与六个钒原子相连的六重桥氧原子，共 3 个。

1992 年，Pettersson 等人用核磁的方法检测到了溶液中存在十三钒氧簇阴离子 $[H_{12}V_{13}O_{40}]^{3-}$，并提出该阴离子具有 Keggin 结构[5]，但没有分离出固体。2005 年，王恩波等人用水热的方法，制出了分子组成为 $(NH_4)_4[H_{12}V_{12}O_{36}(V^{IV}O_4)]\cdot 11H_2O$ 的化合物[6]。化合物阴离子结构如图 11-4 所示。阴离子具有类 Keggin 结构，位于阴离子中心的钒为 +4 价，与之相连的四个氧原子呈无序分布。

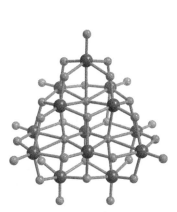

图 11-3　$[V_{13}O_{34}]^{3-}$ 原子连接图

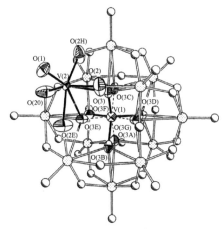

图 11-4　$[H_{12}V_{12}^VO_{36}(V^{IV}O_4)]^{4-}$ 结构

11.1.3　十四钒氧簇

将 $VO_2(acac)$、Et_3N 和 Et_4NCl 在乙腈中反应并在 DMF 中重结晶，可得到

分子组成为 $(Et_4N)_5[V_{14}O_{36}Cl]$ 的化合物[7]，单晶 X 射线分析表明，化合物阴离子由一个 $[V_{14}O_{36}]$ 笼包裹一个 Cl^- 组成（图 11-5）。在十四钒氧簇阴离子 $[V_{14}O_{36}]^{4-}$ 中，所有的钒原子均为五配位四方锥结构。按照键合钒原子数目的不同，阴离子中的氧原子可分为三种：①只与一个钒原子成键的端氧原子，共有 14 个；②与两个钒原子成键的二重桥氧原子，共有 10 个；③与三个钒原子成键的三重桥氧原子，共有 12 个。十四个钒氧四方锥通过共边和共顶点相连构成半开放的笼状结构。笼的表面有两个大的开口，位于开口两侧的原子距离分别为 5.763Å 和 5.744Å。

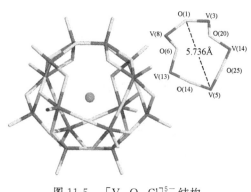

图 11-5　$[V_{14}O_{36}Cl]^{5-}$ 结构

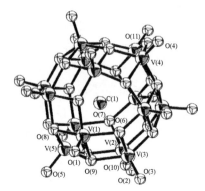

图 11-6　$[V_{15}O_{36}Cl]^{6-}$ 结构

11.1.4　十五钒氧簇

至今，已报道的十五钒氧簇阴离子有三种类型。1987 年，Müller 等人首次报道了含有十五钒氧簇阴离子的化合物 $(Me_4N)_6[V_{15}O_{36}Cl]\cdot 4H_2O$[8]，其阴离子结构如图 11-6 所示。在该化合物中，十五钒氧簇阴离子 $[V_{15}O_{36}]^{5-}$ 呈笼形结构，十五个钒氧四方锥通过共顶点相连形成笼状阴离子，Cl^- 被包在笼的中间。阴离子中的氧原子可分为三种类型：①只与一个钒原子相连的端基氧原子，共 15 个；②与两个钒原子相连的二重桥氧原子，共有 3 个；③与三个钒原子相连的三重桥氧原子，共有 18 个。阴离子中的钒呈 +4 和 +5 两种价态，根据其还原程度的不同，两种价态钒的数目可发生变化。在该化合物阴离子中，共有七个 +5 价的钒，八个 +4 价的钒。

含有同构阴离子的化合物 $[V_{15}O_{36}Cl](NH_4)_3Na_7\cdot 30H_2O$ 中，含有三个 +5 价钒和十二个 +4 价钒[9]。而在化合物 $(Bu^n_4N)_4[V_{15}O_{36}Cl]$ 的阴离子中，则含有九个 +5 价的钒和六个 +4 价的钒[10]。在上述几例化合物中，十五钒氧簇阴离子笼中包的都是氯原子，Yamase 等人通过 $[V_4O_{12}]^{4-}$ 和甲醇在 pH=9（用 K_2CO_3 调 pH 值）的条件下反应，得到了分子组成为 $K_5H_2[V_{15}O_{36}$

(CO_3)]·14.5H_2O 的化合物[11]。在该化合物中，十五钒氧簇阴离子笼中包的是 CO_3^{2-}（图 11-7），碳酸根上的 3 个氧原子通过弱的共价键与阴离子上的钒原子相连。

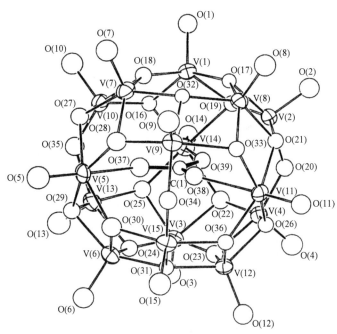

图 11-7 $[V_{15}O_{36}(CO_3)]^{7-}$ 结构

在洪茂春等人报道的化合物（Et_4N）$_5$[$HV_{15}O_{39}$(acac)Cl]·3CH_3CN 中（acac＝乙酰丙酮），含有另一种结构类型的十五钒氧簇阴离子 [$HV_{15}O_{39}$(acac)Cl]$^{4-}$[12]。与上一种十五钒氧簇阴离子相同的是阴离子也含有十五个钒原子，阴离子同样呈笼状结构，而且在笼的中间也包含一个氯离子。与上一种阴离子不同的是在该阴离子中，有一个 acac 配体参与了配位，而且由于 acac 配体的参与，阴离子中有一个钒原子为六配位八面体构型，而且阴离子中的钒均为＋5 价，除 acac 的原子参与配位外，阴离子中还多了三个氧原子，这就使得阴离子中二重桥氧增加，三重桥氧减少，整个阴离子中共有 15 个端氧原子、14 个二重桥氧原子和 10 个三重桥氧原子（图 11-8）。

1993 年，Hill 报道了另一种含十五钒氧簇阴离子的化合物 [TMA]$_3$H$_6$[$V_{15}O_{42}$]·2.5H_2O[13]，其阴离子结构如图 11-9 所示，阴离子中的十五个钒原子呈三种配位构型，位于阴离子中心的钒呈四配位四面体构型，另外有十二个钒原子呈六配位八面体构型，十二个 VO_6 八面体分为四组，每三个 VO_6 八面体通过共边相连形成三金属簇，四组三金属簇通过共顶点相连，形成包裹中心 VO_4 四面体的笼状结构（Keggin 结构），剩余两个钒原子呈五配位四方锥型，分别位于

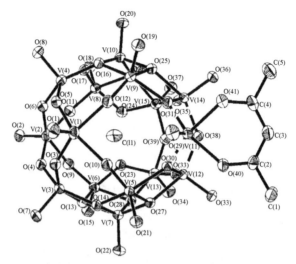

图 11-8 $[HV_{15}O_{39}(acac)Cl]^{4-}$ 结构

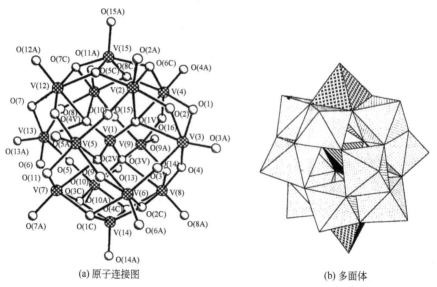

(a) 原子连接图 　　　　　　　　　(b) 多面体

图 11-9 $[V_{15}O_{42}]^{9-}$ 结构

两个由两组三金属簇共顶点相连构成的空洞处，整个阴离子呈双加冠 Keggin 结构。

由 V_2O_5 和吗啉反应，可得到含有十五钒氧簇混合价阴离子的化合物 $[Morp]_6[VO_4V_{14}O_{32}(OH)_6]\cdot 2H_2O^{[14]}$，该化合物阴离子结构与 Hill 报道的化合物 $[TMA]_3H_6[V_{15}O_{42}]\cdot 2.5H_2O$ 中阴离子结构一样，呈双加冠 Keggin 结构，只是阴离子中有三个钒呈 +4 价。由 V_2O_5、NaOH 和巯乙胺反应，同样可得到含有双加冠 Keggin 结构阴离子的化合物 $[C_4N_2S_2H_{14}]_5[H_4V_{15}O_{42}]_2\cdot 10H_2O^{[15]}$。

11.1.5 十六钒氧簇

由 V_2O_5、$Cu(CH_3COOH)_2 \cdot H_2O$、1,2-丙二胺和 $[(CH_3)_4N]OH$ 在水热条件下反应,可得到结构单元为 $[\{Cu(1,2\text{-}pn)_2\}_7\{V_{11}^{IV}V_5^{V}O_{38}(H_2O)\}_2] \cdot 4H_2O$ 的化合物[16],在该化合物中,其阴离子为十六钒氧簇,在该阴离子中,所有的钒原子均呈五配位四方锥构型,阴离子中的氧原子按照键合情况可分为三类:①只与一个钒原子相连的端基氧原子,共 16 个;②与两个钒原子相连的二重桥氧原子,共 2 个;③与三个钒原子相连的三重桥氧原子,共 20 个。整个阴离子中,VO_5 四方锥通过三重桥氧和二重桥氧相连,构成笼状结构,在笼的中间包裹一个水分子,其结构如图 11-10 所示。

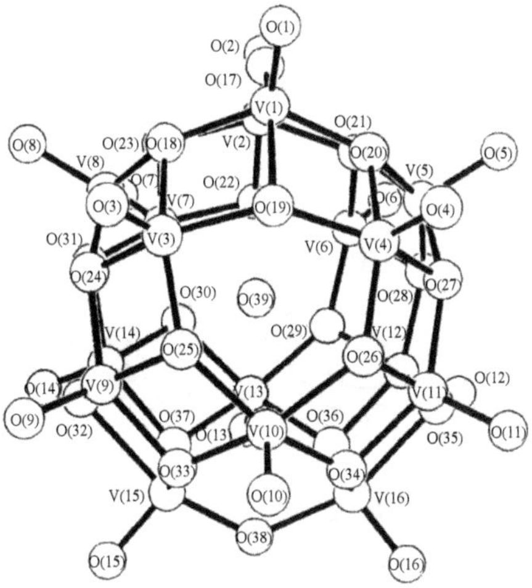

图 11-10 $\{V_{11}^{IV}V_5^{V}O_{38}(H_2O)\}$ 结构

在室温氮气氛下,将 1g $VOSO_4$ 溶于 60mL 水中,向溶液中加入 1.38g $KOH(95\%)$,再加入 1.09g N,N,N',N'-四(2-羟乙基)乙二胺,可得到分子组成为 $K_{10}[H_2V_{16}O_{38}] \cdot 13H_2O$ 的化合物[17],在该化合物中,十六钒氧簇阴离子 $[V_{16}O_{38}]^{12-}$ 与 $[\{Cu(1,2\text{-}pn)_2\}_7\{V_{16}O_{38}(H_2O)\}_2] \cdot 4H_2O$ 中的相同,只是在阴离子笼中未包裹水分子,但值得特别注意的是阴离子中所有的钒均呈 +4 价,即该化合物中十六钒氧簇阴离子的价态为 $[H_2V_{16}^{IV}O_{38}]^{10-}$。$V_2O_5$、$H_2C_2O_4 \cdot 2H_2O$、KOH、$NiCl_2 \cdot 2H_2O$、en 在水热条件下反应,可得到结构单元组成为 $Ni(en)_3\{V_{11}^{IV}V_5^{V}O_{38}Cl[Ni(en)_2]_3\} \cdot 8.5H_2O$ 的化合物。在该化合物中,十六钒氧簇阴离子具有和上述化合物中十六钒氧簇阴离子相同的结构,与上述不同的是阴离子笼的中间包裹 1 个 Cl^-,阴离子中的钒呈混合价态[18]。由 V_2O_5、$CuCl_2 \cdot 2H_2O$ 和 1,2-丙二胺在水热条件下反应,可得到分子组成为 $[Cu(enMe)_2]_{3.5}[V_{16}O_{38}Cl] \cdot 2H_2O$ 的化合物,其阴离子结构和上述十六钒氧簇阴离子结构相同,阴离子笼中间包裹 1 个氯离子。键价计算表明阴离子中钒存在 V^{IV} 和 V^{V} 两种价态,其中 V^{V} 为 6 个,V^{IV} 为 10 个[19],因此,阴离子的准确表达式为 $[V_{10}^{IV}V_6^{V}O_{38}Cl]^{7-}$。$V_2O_5$、$CdCl_2$、1,10-菲啰啉、$H_2C_2O_4 \cdot 2H_2O$、$CH_3COONa$ 和三甲醇氨基甲烷在水热条件下反应,可得到结构单元为 $H_2[C(CH_2OH)_3NH_2]_{0.5}[Cd(phen)_3]\{[Cd(H_2O)(phen)_2]_2(V_{16}O_{38}Cl)\}_{0.5}\{[Cd(H_2O)(phen)_2]_2(V_{16}O_{39}Cl)\}_{0.5} \cdot 2H_2O$

的化合物,在该化合物中,存在 $[V_{16}O_{38}Cl]^{8-}$ 和 $[V_{16}O_{39}Cl]^{9-}$ 两种十六钒氧簇阴离子,其中 $[V_{16}O_{38}Cl]^{8-}$ 阴离子结构与上述十六钒氧簇阴离子结构相同,而 $[V_{16}O_{39}Cl]^{9-}$ 阴离子结构与前述的十六钒氧簇结构略有差别,其中的一个钒原子结合了两个端基氧原子而呈四面体构型[20](图 11-11)。由 V_2O_5、$NaTeO_3$、$Ni(CH_3COO)_2$、4,4'-bpy 和 HCl 在水热条件下反应,可得到结构单元组成为 $[Ni(4,4'-bpy)_2]_2[V_7^{IV}V_9^VO_{38}Cl]\cdot(4,4'-bpy)\cdot 6H_2O$ 的化合物,其十六钒氧簇阴离子结构与上述化合物中的相同[21]。

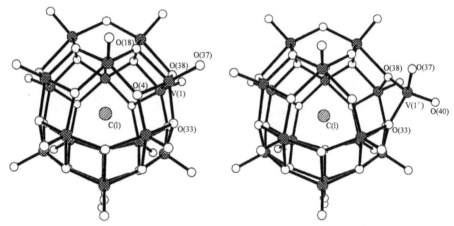

图 11-11 $[V_{16}O_{38}Cl]^{8-}$ 和 $[V_{16}O_{39}Cl]^{9-}$ 阴离子结构

$VO_2(acac)$ 溶于乙腈/DMF 混合溶剂,室温搅拌下加入三乙胺,再加入 $Et_4NBr\cdot 2H_2O$,可得到分子组成为 $(Et_4N)_5[V_{16}O_{40}Br]$ 的化合物;如果将 $Et_4NBr\cdot 2H_2O$ 换为 Et_4NCl,则可生成分子组成为 $(Et_4N)_5[V_{16}O_{40}Cl]$ 的化合物[12]。与上述十六钒氧簇阴离子不同的是,该阴离子中有四十个氧原子,阴离子中所有钒原子均为五配位四方锥构型,十六个 VO_5 四方锥通过共顶点和共边构成十六钒氧簇笼,$Cl^-(Br^-)$ 位于笼的中间(图 11-12),阴离子中的氧原子可分为三类:只与一个钒原子相连的端基氧原子(共 16 个),与两个钒原子相连的二重桥氧原子(共 8 个),与三个钒原子相连的三重桥氧原子(共 16 个)。向 $VO_2(acac)$ 的甲醇/DMF 混合溶液中加入二硫化四甲基硫脲和 $Et_4NClO_4\cdot H_2O$,室温搅拌,可得到分子组成为 $(Me_2NH_2)_8[H_2V_{16}O_{42}(ClO_4)]\cdot 4H_2O$ 的化合物[12]。其阴离子结构如图 11-13 所示,阴离子中十六个钒原子均为五配位四方锥构型,十六个四方锥通过共顶点和共边构成笼状结构,高氯酸根阴离子位于笼的中心,阴离子中共有 16 个端基氧、14 个二重桥氧和 12 个三重桥氧。

在上述的十六钒氧簇阴离子中,尽管其金属原子均为十六个钒原子,但阴离子中的氧原子却有明显差别,阴离子中氧原子个数的不同导致阴离子内氧原子键

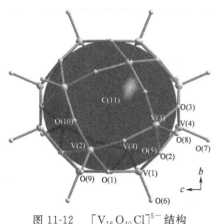 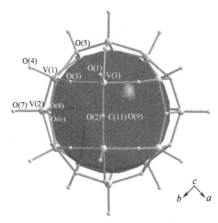

图 11-12　$[V_{16}O_{40}Cl]^{5-}$ 结构　　　　图 11-13　$[H_2V_{16}O_{42}(ClO_4)]^{8-}$ 结构

合方式的改变和阴离子对称性的变化。由十六钒氧簇形成的化合物可知，该阴离子具有较强的配位能力，可以和过渡金属配离子形成多维配位聚合物[16,19-21]；而且，十六钒氧簇阴离子易呈现还原态，最多十六个钒原子均可被还原[17]。

11.1.6　十七钒氧簇

$(Bu_4N)_3[H_3V_{10}O_{28}]$、$Pd(COD)Cl_2$ 和 p-甲基磺酸在乙腈溶液中在氮气氛下回流可制得分子组成为 $(Bu_4N)_4[V_{17}O_{42}]\cdot 2CH_3CN$ 的化合物[22]，在该化合物中，其阴离子为十七钒氧簇（图 11-14），阴离子核心具有八重立方烷结构，

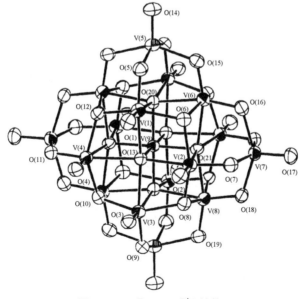

图 11-14　$[V_{17}O_{42}]^{4-}$ 结构

四个 VO_5 四方锥位于八重立方烷的四个面上构成十七钒氧簇阴离子。阴离子中的钒原子存在六配位八面体（13 个）和五配位四方锥（4 个）两种构型；氧原子存在端基氧（12 个）、二重桥氧（16 个）、三重桥氧（8 个）和五重桥氧（6 个）四种类型。阴离子中位于中心的钒和 4 个四方锥构型的钒原子呈 +4 价。

11.1.7　十八钒氧簇

搅拌下向盛有 100mL $CsVO_3$（1.16g）的 90℃水溶液的烧瓶中加入 $180\mu L$

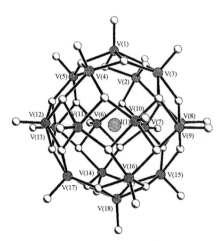

图 11-15　$[H_4V_{18}O_{42}I]^{9-}$ 结构

100%的水合肼，连续搅拌 1min，然后转到 90~95℃的油浴加热 1h，滴加 47%的 HBr 水溶液调体系 pH=7.9~8.0，直到 pH 值稳定，再在 90~95℃的油浴加热 1.5h，5h 内缓慢降温到 20℃，可得到黑色晶体，其分子组成为 $Cs_9[H_4V_{18}O_{42}Br]\cdot 12H_2O$[23]。用 57%的 HI 水溶液代替上述反应中的 HBr 水溶液，可得到分子组成为 $Cs_9[H_4V_{18}O_{42}I]\cdot 12H_2O$ 的化合物，用 KVO_3 代替 $CsVO_3$ 反应，可以得到更稳定的钾盐。化合物中十八钒氧簇阴离子结构如图 11-15 所示。阴离子由十八个 VO_5 四方锥通过 24 个三重桥氧相连而成，每一个钒原子连接一个端基氧原子，端氧位于四方锥的锥顶。十八个钒氧四方锥可分为三组，上下两组各五个四方锥，其中四个四方锥位于近似平面正方形的四个顶点，第五个四方锥位于平面正方形的中间；另八个钒氧四方锥位于阴离子中间，通过共顶点桥氧相连构成四方锥环，八个钒原子几乎位于同一平面。在后续的研究中，又报道了多例包含有不同基团和不同混合价的十八钒氧簇阴离子构成的化合物[17,24-28]。新制得的 $(NH_4)_8[H_9V_{19}O_{50}]\cdot 11H_2O$ 和 NaN_3 与 Et_4NBF_4 在水溶液中反应，经处理后可得到分子组成为 $(Et_4N)_5[H_2V_{18}O_{44}(N_3)]$ 的化合物[29]。在该化合物中，十八钒氧簇阴离子 $[H_2V_{18}O_{44}(N_3)]^{5-}$ 由钒氧四方锥通过共边和共顶点相连形成笼状结构，叠氮酸根阴离子 N_3^- 包裹在笼的中间。由 $[Bu^tNH_3]_4[V_4O_{12}]$ 和叠氮化钠 NaN_3 在水溶液中反应，调节 pH 值，也可得到含十八钒氧簇阴离子的化合物 $Na_{12}H_2[V_{18}O_{44}(N_3)]\cdot 30H_2O$[30]。

11.1.8　十九钒氧簇

8.0g NH_4VO_3 溶于 250mL75℃的水中（pH=5.8），溶液冷却至 70℃，加

入 1.37g 硫酸肼（溶液颜色由黄色变色经橄榄绿到蓝紫色到棕黑色），保温搅拌 5min（pH 值增大到 7.3），搅拌下向溶液中加入 10% 的 H_2SO_4 调 pH 值到 5.3～5.4，然后立即将反应液静置于 65～70℃ 的油浴 5h（用玻璃皿覆盖），可得到分子组成为 $(NH_4)_8[V_{19}O_{41}(OH)_9]\cdot 11H_2O$ 的化合物[31]。

该十九钒氧簇阴离子由 12 个 VO_6 八面体（钒为正四价 V^{IV}）、6 个 VO_4 四面体（钒为正五价 V^V）和 1 个位于中心的 VO_4 四面体构成（钒为正五价 V^V），整个阴离子具有 C_3 对称性，其结构如图 11-16 所示。位于外层的钒原子通过位于阴离子内的 1 个四重氧、4 个三重桥氧和位于壳层的 12 个三重桥氧、15 个二重桥氧原子相连，在 15 个二重桥氧原子中，9 个位于 VO_6 八面体两两之间共用位置的结合 1 个质子，位于 VO_6 八面体和 VO_4 四面体共用位置的 6 个二重桥氧未结合质子。

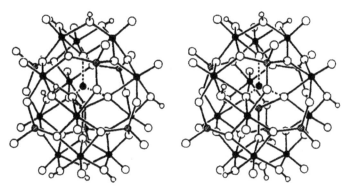

图 11-16　$[V_{19}O_{41}(OH)_9]^{8-}$ 结构

$(n\text{-}C_4H_9)NH_3VO_3$（375mg，2.16mmol）、$VOSO_4\cdot 5H_2O$（253mg，1.1mmol）、$n\text{-}C_4H_9NH_2$（0.2mL，2.02mmol）和 H_2O（5mL）加热回流 3h，反应液过滤，冷却过夜，可得到分子组成为 $C_{36}H_{108}N_9O_{49}V_{19}\cdot 7H_2O$ 的棕黑色块状晶体。$(n\text{-}C_3H_7NH_3)_9[V_{19}O_{49}]\cdot 7H_2O$ 可用同样的方法制备[32]。化合物晶体结构分析表明，其阴离子结构为十九钒氧簇（图 11-17）。其阴离子对称性与上述 $[V_{19}O_{41}(OH)_9]^{8-}$ 对称性相同，也为 C_3 对称性，与上述阴离子不同的是，在该阴离子中，构成笼的有 12 个 VO_6 八面体、3 个 VO_5 四方锥和 3 个 VO_4 四面体。与前述阴离子不同之处还在于在该阴离子中 V^{IV} 比较少，从而表现为前述阴离子为 -8 价，且阴离子中有 9 个质子化的氧，而在该化合物中，阴离子为 -9 价，阴离子中没有质子化的氧。

3mL $LiOH\cdot H_2O$（5mmol）水溶液加入 10mL 95℃ 的 V_2O_5（2.5mmol）浆状液中，用固体硫酸肼（2.5mmol）处理溶液，再加热 5～10min，将该黑色溶液稀释到 25mL，接着加入 $KMnO_4$（1.25mmol）并加热 1.5h，过滤后在室温放

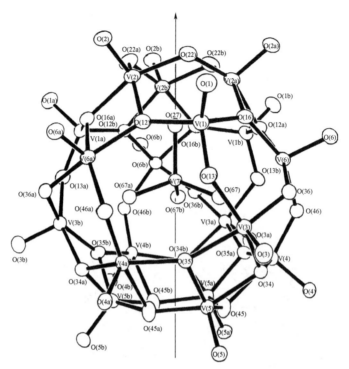

图 11-17　$[V_{19}O_{49}]^{9-}$ 结构

置，得到结构单元组成为 $[H_6Mn_3V_{15}^{IV}V_4^VO_{46}(H_2O)_{12}]\cdot 3H_2O$ 的化合物[33]，化合物基本构筑块为十九钒氧簇 $[V_{19}O_{46}]^{12-}$。该十九钒氧簇阴离子 $[V_{19}O_{46}]^{12-}$ 的结构可看作是在十八钒氧簇基团 $[V_{18}O_{42}]$ 中间包裹了 1 个 $[VO_4]^{3-}$ 四面体阴离子，$[VO_4]^{3-}$ 四面体中的每 1 个氧原子为均四重桥氧，与簇壳层上的 3 个 $[VO_6]$ 八面体中的钒原子相连，$[V_{19}O_{46}]^{12-}$ 中的 12 个 $[VO_6]$ 八面体通过共棱和 6 个 $[VO_5]$ 四方锥相连。每个 $[VO_6]$ 八面体连接 1 个端基氧、4 个三重桥氧和 1 个四重桥氧。每个 $[VO_5]$ 四方锥连接 4 个三重桥氧和 1 个端基氧。整个阴离子中氧原子可分为三类：18 个端基氧，24 个三重桥氧和 4 个四重桥氧原子。

11.1.9　二十二钒氧簇

250mL 含有 Et_4NClO_4 (16.0g，69.65mmol) 的水溶液用 6.6g(3.11mmol) 新制的 $(NH_4)_8[H_9V_{19}O_{50}]\cdot 11H_2O$ 处理，转移到 500mL 烧杯中，混合物在不断搅拌下 75℃ 加热 60h，过滤析出的黑色晶体，得到分子组成为 $(NEt_4)_6[HV_{22}O_{54}(ClO_4)]$ 的化合物[29]。在该化合物中，阴离子簇内 VO_5 四方锥通过共边和共顶点相连，高氯酸根作为客体存在于由 VO_5 四方锥形成的 $[HV_8^{IV}V_{14}^VO_{54}]^{5-}$

簇中。

11.1.10 三十四钒氧簇

将偏钒酸钾（3.45g，25.0mmol）溶于 50mL 90℃ 的热水中，清液在 90℃ 下加入 182μL 水合肼（100%，3.75mmol），保温 1h 后，用冰醋酸调溶液 pH 值到 3.8，再保温 2h 热滤，滤液在 90℃ 保温 3h 后冷却至室温，放置得到黑色针状晶体，其分子组成为 $K_{10}[V_{16}^{IV}V_{18}^{V}O_{82}] \cdot 20H_2O^{[34]}$。

晶体结构分析表明（图 11-18），三十四钒氧簇阴离子 $[V_{34}O_{82}]^{10-}$ 具有近似 D_{2d} 对称性，由 30 个 VO_5 四方锥构成的椭球形鞘和中心的 $\{V_4O_4\}O_4$ 立方体构成。

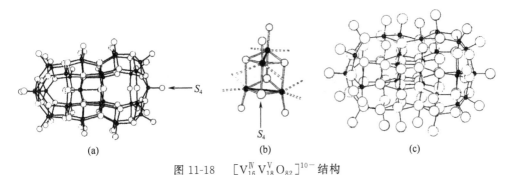

图 11-18　$[V_{16}^{IV}V_{18}^{V}O_{82}]^{10-}$ 结构

11.2　铌氧簇

11.2.1　二十铌氧簇

将 $TBA_4H_4[Nb_6O_{19}] \cdot 7H_2O$(1.0g，0.51mmol) 溶于 20mL 四氢呋喃中，以 100mL/min 的流速向溶液中通入 NO 气体 6min，过滤掉少量固体，向每毫升滤液中加入 0.8mL 乙酸乙酯，可得到分子组成为 $TBA_8[Nb_{20}O_{54}] \cdot H_2O$ 的化合物[35]。

晶体结构分析表明，二十铌氧簇阴离子由两个十铌氧簇阴离子缩合而成（图 11-19）。

11.2.2　二十四铌氧簇

$Rb_8Nb_6O_{19} \cdot 14H_2O$(0.95g，0.5mmol)、$Cu(NO_3)_2 \cdot 2.5H_2O$(0.48g，2.0mmol)、乙二胺（10g，167mmol）和水（10g，550mmol）在 50mL 烧杯中混合，60℃ 搅拌 30min，放置可得到分子组成为 $(H_2en)[Cu(en)_2(H_2O)_2]_3[(\{Nb_{24}O_{72}H_9\}\{Cu(en)_2(H_2O)\}_2\{Cu(en)_2\})_2]$ 的化合物[36]。改变反应原料，用

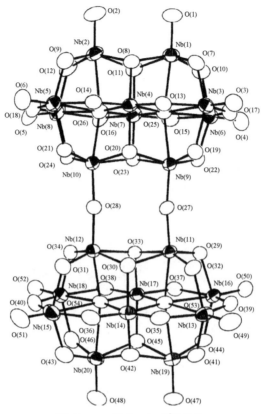

图 11-19 $[Nb_{20}O_{54}]^{8-}$ 结构

铌酸钾 $Rb_8Nb_6O_{19} \cdot 14H_2O$ 代替也可以得到含同构铌氧簇阴离子的化合物[37,38]。

晶体结构分析表明，二十四铌氧簇阴离子是由三个 Nb_7O_{22} 基团和三个 NbO_6 八面体交替相连形成的环状簇（图 11-20）。

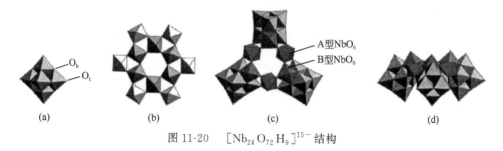

图 11-20 $[Nb_{24}O_{72}H_9]^{15-}$ 结构

11.2.3 二十七铌氧簇和三十一铌氧簇

$K_7HNb_6O_{19} \cdot 13H_2O$（1.0g，0.73mmol）和二苄基二硫代氨基甲酸钠（0.4g，1.4mmol）溶于 8mL 水中，置于反应釜内，200℃下反应 3d 后冷至室

温,过滤掉白色无定形杂质,根据结晶条件的不同,母液中可分别析出二十七铌氧簇和三十一铌氧簇。室温下向母液中扩散甲醇,可得到组成为 $K_{13}Na_3[HNb_{27}O_{76}]\cdot25H_2O$ 的化合物及六铌氧簇晶体。室温下缓慢蒸发母液可得到分子组成为 $K_{19}Na_4[H_{10}Nb_{31}O_{93}(CO_3)]\cdot35H_2O$ 的化合物晶体及六铌氧簇晶体[39]。阴离子晶体结构如图 11-21 和图 11-22 所示。

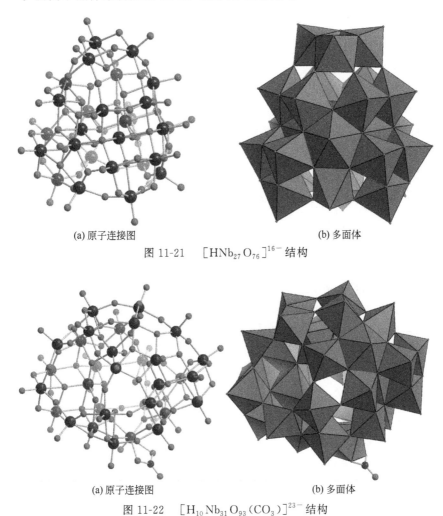

(a) 原子连接图　　　　　(b) 多面体

图 11-21　$[HNb_{27}O_{76}]^{16-}$ 结构

(a) 原子连接图　　　　　(b) 多面体

图 11-22　$[H_{10}Nb_{31}O_{93}(CO_3)]^{23-}$ 结构

11.3　钼氧簇

11.3.1　十三钼氧簇

1992 年,Klemperer 等人用 $MoO_3\cdot2H_2O$(83mg,0.46mmol) 和 $[(C_5Me_5Rh)_2$

$(OH)_3]Cl(0.10g,0.16mmol)$ 在 2mL 水中在密闭容器内 190℃ 反应 2d，得到分子组成为 $[(C_5Me_5Rh)_8(Mo_{13}O_{40})]Cl_2 \cdot 13H_2O$ 的化合物[40]，在该化合物中存在一个 $[(C_5Me_5Rh)_8(Mo_{13}O_{40})]^{2+}$，该阳离子由 1 个 $[Mo_{13}O_{40}]^{14-}$ 和 8 个 $[(C_5Me_5)Rh]^{2+}$ 基团组成，阴离子和 $[(C_5Me_5)Rh]^{2+}$ 基团之间通过阴离子中的 24 个桥氧相连，阴离子由 12 个 Mo^V 构成的钼氧八面体和一个位于中心的 Mo^{VI} 构成的钼氧四面体组成。13 个钼和 40 个氧构成具有 Keggin 结构特征的阴离子，由于阴离子中位于多原子位置的所有钼原子均被还原为五价，在一些钼原子间存在金属键（2.661Å）。

将 $(NH_4)_6Mo_7O_{24} \cdot 4H_2O(177mg,0.14mmol)$、$NH_2OH \cdot HCl(700mg,10mmol)$ 和 $Bu_4NBr(500mg,1.5mmol)$ 在氮气氛下 90℃ 反应 10h，反应混合物变为深绿色，用丙酮萃取深绿色产物，真空干燥，在 CH_2Cl_2-MeOH 混合溶剂中重结晶，得到深绿色晶体，其分子组成为 $[Bu_4N]_6[H_3O]_2[Mo_{13}O_{40}]_2$，晶体结构分析表明，分子中含有两个不同对称性的阴离子 $[Mo_{13}O_{40}]^{2-}$，这两个阴离子具有 Keggin 结构，中心原子为钼[41]。

11.3.2 十六钼氧簇

$(NH_4)_6Mo_7O_{24} \cdot 4H_2O(3.18g,2.57mmol)$ 和苯基膦酸（0.6g,3.80mmol）溶于 100mL 85℃ 的水中，加入盐酸肼（0.3g,2.86mmol）后，溶液颜色迅速由黄变绿，最后至深蓝色。85℃ 下再搅拌 15min，得到红色溶液，用氯化二甲铵溶液（1.65g,20.23mmol 溶于 DMF/水为 50mL/10mL）处理，在 75℃ 下加热，将反应混合液体积减至 70mL，20℃ 放置 12h，可得到分子式为 $(Me_2NH_2)_6$$[H_2Mo_{16}(OH)_{12}O_{40}]$ 的化合物[42]。

将 MoO_3、$Na_2MoO_4 \cdot 2H_2O$、$C(CH_2OH)_4$、$(Et_4N)Cl$、Me_3NH 和水按摩尔比 6:6:10:10:10:300 混合，在 160℃ 下反应 3d，可得到分子组成为 $(Me_3NH)_2(Et_4N)Na_4[Na(H_2O)_3H_{15}Mo_{42}O_{109}\{(OCH_2)CCH_2OH\}_7] \cdot 15H_2O$ 的红棕色晶体和红棕色溶液，滤去固体，放置滤液，可得到分子组成为 $(NH_4)_7[NaMo_{16}(OH)_{12}O_{40}] \cdot 4H_2O$ 的化合物[42]。改变反应条件，可得到含有相同阴离子的化合物 $(NH_4)_7[NaMo_{16}(OH)_{12}O_{40}] \cdot 4H_2O$ 和 $(Me_3NH)_4K_2$$[H_2Mo_{16}(OH)_{12}O_{40}] \cdot 8H_2O$[43]。晶体结构分析表明，上述化合物中存在一个十六钼氧簇 $[Mo_{16}(OH)_{12}O_{40}]^{8-}$（图 11-23）。在该阴离子的中心有一个空腔，里面可以包含质子或 Na^+。阴离子中有 4 个 Mo^{VI} 中心和 12 个 Mo^V 中心，40 个氧原子和 12 个羟基，其中的 12 个 Mo^V 中心两两形成双核单元，双核单元内的 Mo—Mo 距离为 2.62(1) Å，具有单键性质。

六亚甲基四胺（2.2g,15.7mmol）室温下溶于 20mL 水中，用 1.2mL

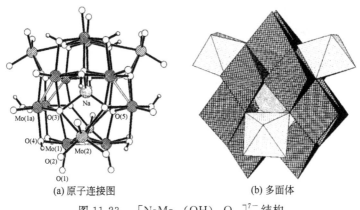

(a) 原子连接图 (b) 多面体

图 11-23　$[NaMo_{16}(OH)_{12}O_{40}]^{7-}$ 结构

(37%) 的盐酸酸化所得溶液，搅拌下向酸化后的溶液中同时加入 $Na_2MoO_4 \cdot 2H_2O$ (0.96g, 4.0mmol) 和 $Na_2S_2O_4$ (0.07g, 0.44mmol)，加入后溶液的颜色首先变为绿色，然后变为黄绿色，最后变为棕色，溶液 pH 值为 4.0~4.5，溶液过滤放置，得到分子组成为 $[C_6H_{13}N_4]_{10}[H_2Mo_{16}O_{52}] \cdot 34H_2O$ 的棕色晶体[44]。晶体结构分析表明，化合物的阴离子呈平面型（图 11-24）。氧化还原滴定，键价计算和元素分析表明阴离子中的钼呈两种价态，其中 12 个钼为 Mo^{VI}，4 个钼为 Mo^V，其阴离子为 $[H_2Mo_4^V Mo_{12}^{VI}O_{52}]^{10-}$，五价的钼两两构成双核单元，两个钼原子间的距离为 2.6427(4)Å，在 Mo—Mo 单键的范围之内，这也是化合物呈棕色的原因[45]。

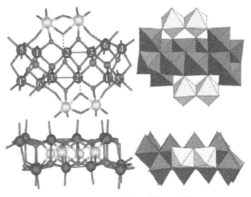

图 11-24　$[H_2Mo_4^V Mo_{12}^{VI}O_{52}]^{10-}$ 结构

11.3.3　三十六钼氧簇

1973 年，Glemser 报道在酸化钼酸盐时，在溶液或固体盐中存在一个前所未有大的同多阴离子，并通过拉曼光谱、分子量测定、X 射线单晶分析等一系列手段确定该阴离子的组成为 $Mo_{36}O_{112}^{8-}$[46]。Paulat-Böschen 通过单晶 X 射线衍射确定了三十六钼氧簇钾盐 $K_8[Mo_{36}O_{112}(H_2O)_{16}] \cdot 36H_2O$ 的结构[47,48],该阴离子的结构与 Glemser 提出的结构仅存在配位水的差别。三十六钼氧簇阴离子 $[Mo_{36}O_{112}(H_2O)_{16}]^{8-}$ 由两个十八钼氧簇亚单元构成,两个亚单元互成中心对称,通过四个共用氧原子连接,构成一个环,一些阳离子和水分子位于环的中央,其结构如图 11-25 所示。值得注意的是,在三十六钼氧簇中水分子参与了配位,其中有四个钼为七配位构型。

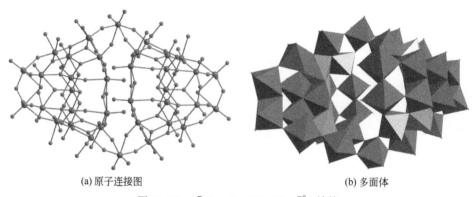

(a) 原子连接图　　　　　　　　　　(b) 多面体

图 11-25　$[Mo_{36}O_{112}(H_2O)_{16}]^{8-}$ 结构

1986 年章士伟等人用 $Na_2MoO_4 \cdot 2H_2O$ 和 $NH_2OH \cdot HCl$ 在酸性水溶液中回流,得到了分子组成为 $[Mo_{36}O_{110}(NO)_4(H_2O)_{14}]$ 的化合物[49],与前述三十六钼氧簇不同的是,其中四个亚硝基取代了 4 个与钼配位的水。在后续的研究工作中,人们陆续得到了以上述三十六钼氧簇为核心的多个化合物,$[Mo_{36}O_{108}(NO)_4(MoO)_2La_2(H_2O)_{28}]_n \cdot 56nH_2O$[50],$(NH_4)_4Na_2[Mo_{36}^{VI}O_{108}(NH_2OH)_2(OH)_6(H_2O)_{12}] \cdot 35H_2O$[51],$[\{Gd(H_2O)_5\}_4\{Mo_{36}(NO)_4O_{108}(H_2O)_{16}\}] \cdot 34H_2O$[52],$(H_3O)_2[\{La(H_2O)_5\}_2\{La(H_2O)_6\}\{La(H_2O)_5Cl\}\{Mo_{36}(NO)_4O_{108}(H_2O)_{16}\}] \cdot Cl \cdot 2H_2O$,$[\{Ln(H_2O)_6\}_4\{Ln(H_2O)_4\}\{Mo_{36}(NO)_4O_{108}(H_2O)_{16}\}]Cl_3 \cdot nH_2O$ (Ln = Nd,Sm),$[\{Ln(H_2O)_6\}_2\{Ln(H_2O)_7\}_2\{Mo_{36}(NO)_4O_{108}(H_2O)_{16}\}] \cdot nH_2O$(Ln=La,Ce,Pr,Nd),$(H_3O)_3[\{Ln(H_2O)_6\}_2\{La(H_2O)_4\}\{Mo_{36}(NO)_4O_{108}(H_2O)_{16}\}] \cdot nH_2O$(Ln=Tb,Dy,Ho,Er,Yb,Lu)[53],尽管其形成条件和分子式多不相同,但其化合物的核心结构三十六钼氧簇阴离子相同。

11.3.4　三十七钼氧簇

将含有 0.3g $[NH_3Pr]_4[Mo_8O_{26}]$ 和 2mL 甲醇的 25mL 水溶液(pH=3.3)置于玻璃管中,在氮气氛中用 500W 超高压汞灯光照 20h,pH 值变为 4.4,加

入 2g 四甲基氯化铵固体，4℃下放置 10～14d，可得到红棕色晶体。元素分析和晶体结构测定表明，化合物分子式为 $[Me_4N]_{12}[H_{14}Mo_{37}O_{112}] \cdot 16H_2O^{[54]}$。在该化合物中，存在一个三十七钼氧簇阴离子。晶体结构分析表明，$[H_{14}Mo_{24}^VMo_{13}O_{112}]^{12-}$ 具有夹心型结构，$[H_{10}Mo_{12}^VO_{40}(Mo^{VI}O_2)_3]^{4-}$ 位于夹层部分，两个 $[H_2Mo_6^VMo_5^{VI}O_{33}]^{4-}$ 位于阴离子的两端（图 11-26），20 个五价的钼形成 10 个 Mo—Mo 单键，其键长为 2.56(1)～2.625(7) Å。

11.3.5　四十二钼氧簇

由 MoO_3、$Na_2MoO_4 \cdot 2H_2O$、$C(CH_2OH)_4$、Et_4NCl、Me_3NHCl 和水按摩尔比 6：6：10：10：10：300 在 160℃下反应 3d，可得到分子

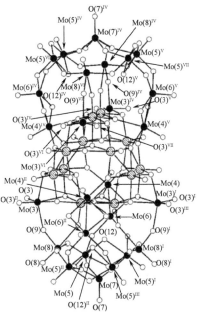

图 11-26　$[H_{14}Mo_{37}O_{112}]^{12-}$ 结构

组成为 $(Me_3NH)_2(Et_4N)Na_4[Na(H_2O)_3H_{15}Mo_{42}O_{109}\{(OCH_2)_3CCH_2OH\}_7] \cdot 15H_2O$ 的红棕色晶体。晶体结构测定表明，化合物分子中存在一个四十二钼氧簇阴离子 $[Na(H_2O)_3H_{15}Mo_{42}O_{109}\{(OCH_2)_3CCH_2OH\}_7]^{7-[55]}$。

11.3.6　四十六钼氧簇

将钼酸钠和 $CH_3COONa \cdot 3H_2O$ 反应，通过控制反应条件，可以得到分子组成为 $Na_{21}\{[Na_5(H_2O)_{14}] \subset [Mo_{46}O_{134}(OH)_{10}(\mu\text{-}CH_3COO)_4]\} \cdot CH_3COONa \cdot 95H_2O$ 的化合物，晶体结构分析表明，化合物中存在一个四十六钼氧簇 $\{[Na_5(H_2O)_{14}][Mo_{20}^VMo_{26}^{VI}O_{134}(OH)_{10}(CH_3COO)_4]\}^{21-[56]}$，阴离子结构如图 11-27 所示。

11.3.7　五十四钼氧簇

向 40 mmol 钼酸钠和 50mmol $CH_3COONa \cdot 3H_2O$ 的 50mL 水溶液中加入 8.5mL 17% 的盐酸酸化 pH=4.15，搅拌下 5min 内加入 $NH_2NH_2 \cdot 2HCl$（4.8mmol，0.5g），得到的反应混合物室温放置 10d，颜色由绿色变为深棕色，得到深红色粒状晶体，元素分析和晶体结构测定表明化合物组成为 $Na_{26}[\{Na(H_2O)_2\}_6\{(\mu_3\text{-}OH)_4Mo_{20}^VMo_{34}^{VI}O_{164}(\mu_2\text{-}CH_3COO)_4\}] \cdot \sim 120H_2O^{[57]}$，阴离

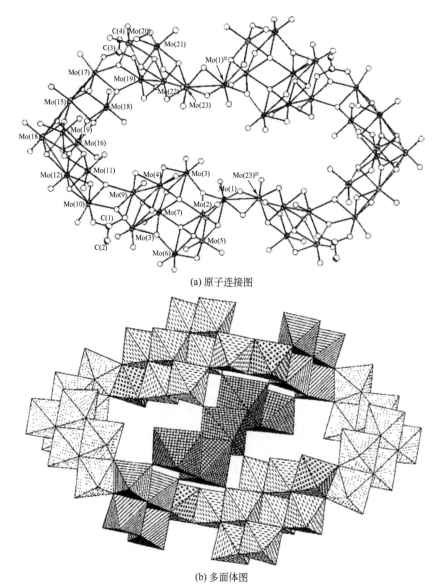

(a) 原子连接图

(b) 多面体图

图 11-27 $[Mo_{20}^{V}Mo_{26}^{V}O_{134}(OH)_{10}(CH_3COO)_4]^{26-}$ 原子连接图和 $\{[Na_5(H_2O)_{14}][Mo_{20}^{V}Mo_{26}^{VI}O_{134}(OH)_{10}(CH_3COO)_4]_2\}^{21-}$ 多面体图

子结构如图 11-28 所示。

11.3.8 其他高核钼氧簇

$EuCl_3 \cdot 6H_2O$ 和 K_2MoO_4 反应，控制反应条件并进行一系列的处理，可得到含有一百二十八钼氧簇阴离子 $[\{Mo_{128}Eu_4O_{388}H_{10}(H_2O)_{81}\}_2]^{20-}$ 的化合物[58]。

七钼酸铵 $(NH_4)_6Mo_7O_{24}$ 和醋酸铵 CH_3COONH_4 在 $N_2H_4 \cdot H_2SO_4$ 存在

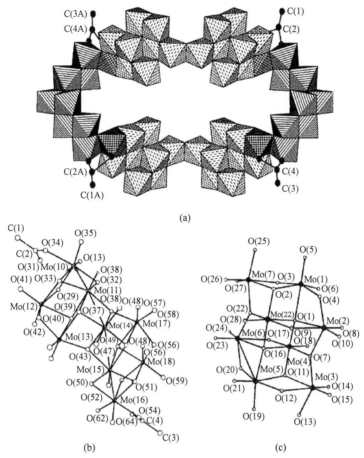

图 11-28 $[(\mu_3\text{-OH})_4\text{Mo}_{20}^{V}\text{Mo}_{34}^{VI}\text{O}_{164}(\mu_2\text{-CH}_3\text{COO})_4]^{32-}$ 结构

下控制条件反应，通过一系列的再处理，可得到分子组成为 $(\text{NH}_4)_{42}$ $[\text{Mo}_{72}^{VI}\text{Mo}_{60}^{V}\text{O}_{372}(\text{CH}_3\text{COO})_{30}(\text{H}_2\text{O})_{72}] \cdot \sim 300\text{H}_2\text{O} \cdot \sim 10\text{CH}_3\text{COONH}_4$ 的化合物[59]。钼酸钠 $\text{Na}_2\text{MoO}_4 \cdot 2\text{H}_2\text{O}$、$\text{Na}_2\text{SO}_4 \cdot 10\text{H}_2\text{O}$ 和硫酸肼在稀硫酸中反应，可得到分子组成为 $\text{Na}_{26}[\text{Mo}_{142}\text{O}_{432}(\text{H}_2\text{O})_{58}\text{H}_{14}] \cdot \sim 300\text{H}_2\text{O}$ 的化合物[60]。从 $\text{Na}_8[\text{Mo}_{36}\text{O}_{112}(\text{H}_2\text{O})_{16}] \cdot 58\text{H}_2\text{O}$ 出发，也可制得由一百四十二钼氧簇构成的化合物[61]。钼酸钠 $\text{Na}_2\text{MoO}_4 \cdot 2\text{H}_2\text{O}$ 在酸性溶液中用铁粉还原，控制反应条件，可制得分子式为 $\text{Na}_{15}[\text{Mo}_{144}\text{O}_{409}(\text{OH})_{28}(\text{H}_2\text{O})_{56}] \cdot \sim 250\text{H}_2\text{O}$ 的化合物[62]。钼酸钠溶液用盐酸酸化后加入硫代硫酸钠并加入少量甲酸，可得到分子组成为 $\text{Na}_{22}[\text{Mo}_{118}^{VI}\text{Mo}_{28}^{V}\text{O}_{442}\text{H}_{14}(\text{H}_2\text{O})_{58}] \cdot \sim 250\text{H}_2\text{O}$ 的化合物。其中包含一个由 146 个钼原子构成的混合价钼氧簇[63]。

钼酸钠溶液酸化后用氯化亚锡 $\text{SnCl}_2 \cdot 2\text{H}_2\text{O}$ 还原，可得到分子组成为 Na_{16} $[\text{Mo}_{152}\text{O}_{429}(\mu_3\text{-O})_{28}\text{H}_{14}(\text{H}_2\text{O})_{66.5}] \cdot \sim 300\text{H}_2\text{O}$ 的化合物[64]，化合物分子内包

含一个由152个钼原子构成的钼氧簇阴离子。钼酸钠溶液酸化后用硫代硫酸钠 $Na_2S_2O_3$ 还原,可得到分子组成为 $Na_{15}[Mo^{VI}_{126}Mo^{V}_{28}O_{462}H_{14}(H_2O)_{70}]_{0.5}[Mo^{VI}_{124}Mo^{V}_{28}O_{457}H_{14}(H_2O)_{68}]_{0.5} \cdot \sim 250H_2O$ 的化合物,其中也包含一百五十二钼氧簇阴离子[63]。

$Na_2MoO_4 \cdot 2H_2O$、NH_4VO_3、$NH_2OH \cdot HCl$ 和盐酸在水溶液中反应,可得到分子组成为 $(NH_4)_{25\pm 5}[Mo_{154}(NO)_{14}O_{420}(OH)_{28}(H_2O)_{70}] \cdot \sim 350H_2O$ 的化合物,其中包含1个由154个钼原子构成的钼氧簇离子[65]。$[Mo_{154}(NO)_{14}O_{420}(OH)_{28}(H_2O)_{70}]^{(25\pm 5)-}$ 中的14个 $[MoNO]^{3+}$ 基团被 $[MoO]^{4+}$ 基团取代后,则形成组成为 $[(MoO_3)_{154}(H_2O)_{70}H_x]^{y-}$ 的钼氧簇阴离子,该阴离子就是早已发表并广泛讨论过的钼蓝的基本结构[66]。控制反应条件,用硫代硫酸钠还原钼酸钠,在所得到的化合物 $Na_{15}[Mo^{VI}_{126}Mo^{V}_{28}O_{462}H_{14}(H_2O)_{70}]_{0.5}[Mo^{VI}_{124}Mo^{V}_{28}O_{457}H_{14}(H_2O)_{68}]_{0.5} \cdot \sim 400H_2O$ 中,即存在一百五十二钼氧簇阴离子,也存在一百五十四钼氧簇阴离子[63]。

将七钼酸铵溶于水,用盐酸调节溶液的酸度到不同的 pH 值,再加入盐酸羟胺,可以形成一系列由一百五十四钼氧簇形成的降解产物[67](表11-1),表明一百五十四钼氧簇的结构对溶液酸度非常敏感。

表 11-1 溶液条件与晶体结构

	母液 pH 值	Mo/O	预测簇结构	簇间连接
1	4.5~3.0	138/476	$[Mo_{138}O_{410}(OH)_{20}(H_2O)_{46}]^{40-}$	分立
2	3.3	138/468	$[Mo_{138}O_{410}(OH)_{20}(H_2O)_{38}]^{40-}$	2维
3	3.0	138/468	$[Mo_{138}O_{406}(OH)_{16}(H_2O)_{46}]^{28-}$	2维
4	3.3~2.5	142/490	$[Mo_{142}O_{400}(OH)_{52}(H_2O)_{38}]^{28-}$	分立
5	2.2	142/490	$[Mo_{142}O_{432}(H_2O)_{58}]^{40-}$	分立
6	1.7	148/507	$[Mo_{148}O_{436}(OH)_{15}(H_2O)_{56}]^{27-}$	3维
7	1.4	150/517	$[Mo_{150}O_{451}(OH)_{5}(H_2O)_{61}]^{35-}$	1维
8	1.4	150/518	$[Mo_{150}O_{442.5}(OH)_{11.5}(H_2O)_{64}]^{24.5-}$	分立{Mo_{150}}
		152/520	$[Mo_{152}O_{446}(OH)_{20}(H_2O)_{54}]^{28-}$	1维{Mo_{152}}

钼酸锂(Li_2MoO_4)溶于水,盐酸酸化后加入氯化亚锡($SnCl_2 \cdot 2H_2O$),可制得分子组成为 $[(MoO_3)_{176}(H_2O)_{80}H_{32}] \cdot \sim 400H_2O \cdot \sim 20Li \cdot \sim 20Cl$ 的化合物,其中包含一个由176个钼原子构成的钼氧簇[68]。钼酸钠 $Na_2MoO_4 \cdot 2H_2O$ 溶于水,酸化后加入铁粉,也可得含有176个钼的化合物 $[Mo_{176}O_{496}(OH)_{32}(H_2O)_{80}] \cdot (600\pm 50)H_2O$[69]。

钼酸钠溶液酸化后用盐酸肼的甲酸溶液还原，可得到分子组成为 Na_{16} $[(MoO_3)_{176}(H_2O)_{63}(CH_3OH)_{17}H_{16}]\cdot\sim600H_2O\cdot\sim6CH_3OH$ 的化合物，其中包含一个由 176 个钼原子构成的钼氧簇[60]。

$Na_{0.4}Co_{0.8}(MoO_4)\cdot1.5H_2O$ 悬浊液酸化后与铁粉反应，可得到分子组成为 $Na_{15}Fe_3Co_{16}[Mo_{176}O_{528}H_3(H_2O)_{80}]Cl_{27}\cdot450H_2O$ 的化合物[70]。研究表明，阴离子与表面的配位水可以被甲醇分子取代[71]。一百七十六钼氧簇阴离子在溶液中保存时，如果有过量的钼酸根离子存在，可以继续长大，成为包含有 248 个钼原子的簇[72]。

连二亚硫酸钠作为还原剂加入用硫酸酸化后的钼酸钠水溶液中，可制得分子组成为 $Na_{48}[H_xMo_{368}O_{1032}(H_2O)_{240}(SO_4)_{48}]\cdot\sim1000H_2O$ 的化合物，其核心为由 368 个钼原子构成的钼氧簇阴离子，这是迄今为止报道的最大的钼氧簇阴离子[73]。

11.4 钨氧簇

11.4.1 十一钨氧簇

由十一钨氧簇阴离子独立构成的化合物目前尚未见于文献，但由十一钨氧簇阴离子参与形成的阴离子已有报道。在化合物 $Na_{12}H[(W_5O_{18})Tb(H_2W_{11}O_{39})]\cdot42H_2O$ 中，单缺位偏钨酸阴离子与 Tb 配位，参与阴离子 $[(W_5O_{18})Tb(H_2W_{11}O_{39})]^{13-}$ 的形成[74]。

11.4.2 十二钨氧簇

十二钨氧簇是目前报道得最多的钨氧簇阴离子，由于阴离子氧原子数和结构的不同，十二钨氧簇阴离子分为偏钨酸阴离子和仲钨酸阴离子。

(1) 偏钨酸阴离子

控制酸化钨酸钠水溶液，即可形成偏钨酸根 $[H_2W_{12}O_{40}]^{6-}$，由钨酸和钨酸钠在一定条件下聚合，可以很高产率地制得偏钨酸钠 $Na_6[H_2W_{12}O_{40}]\cdot nH_2O^{[75]}$。偏钨酸阴离子 $[H_2W_{12}O_{40}]^{6-}$ 的结构首次由 Signer 通过 X 射线粉末衍射测得[76]，具有与 Keggin 结构阴离子相同的结构，其中的两个氢位于阴离子的空腔中，起着和 Keggin 结构中的杂原子同样的作用，核磁共振对此提供了实验佐证[77,78]，单晶结构分析同样表明偏钨酸阴离子和 Keggin 结构阴离子同构[79-89] (图 11-29)。

偏钨酸根易于得到电子被还原形成稳定的还原产物，最多单个阴离子可以得到 32 个电子而结构不发生变化[82]。六电子还原产物 $Rb_4H_8[H_2W_{12}O_{40}]\cdot18H_2O$ 和 $[NH_4]_4H_8[H_2W_{12}O_{40}]$ 的晶体结构研究表明[83,84]，在还原产物中，

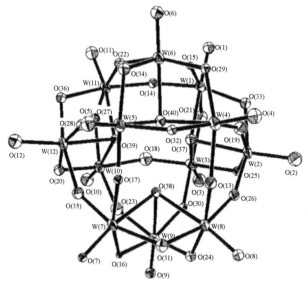

图 11-29 $[H_2W_{12}O_{40}]^{6-}$ 结构

阴离子中的一组三金属簇 $W_3^{VI}O_{13}$ 中的 W^{VI} 被还原为 W^{IV}，形成 $W_3^{IV}O_{13}$ 金属簇，并且在 $W_3^{VI}O_{13}$ 金属簇中存在 W—W 金属键[83]。偏钨酸根阴离子可以作为电子受体与含氮有机给体结合，形成具有超分子作用的化合物[84-86]。

由于偏钨酸阴离子高的负电荷和大量的表面氧原子，可以和过渡金属配离子形成具有分立结构的化合物[87-89]，也可以通过端基氧原子或桥氧原子与过渡金属配离子配位，形成具有担载结构的化合物[90-96]。

偏钨酸根阴离子与稀土配离子结合，可以形成具有一维、二维和三维结构的配位聚合物[97-103]。

(2) 仲钨酸阴离子

偏钨酸和仲钨酸在溶液中可以相互转化，当用酸电位滴定碱金属钨酸盐的水溶液时，在 $H^+/WO_4^{2-}=1.1\sim1.2$ 和 1.5 时得到两个 pH 拐点，第一个拐点相当于形成仲钨酸盐，第二个拐点相当于形成偏钨酸盐。

仲钨酸阴离子 $[H_2W_{12}O_{42}]^{10-}$ 结构早在 20 世纪 70 年代就有多例报道[104-108]，结构如图 11-30 所示。1983 年，Evans 利用中子衍射确定了阴离子中两个质子的位置，两个质子分别与两组三金属簇的共用氧原子相连[109]，此后，又有多例仲钨酸阴离子结构的报道[110-112]，并有文献报道了偏钨酸阴离子 $[H_2W_{12}O_{42}]^{10-}$ 的晶体结构[112]。

以仲钨酸根阴离子为构筑块，与金属离子或配离子配位，可以形成具有多维结构的配位聚合物[113-117]。

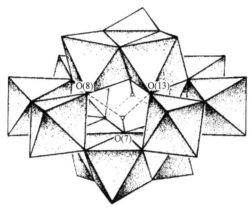

图 11-30　$[H_2W_{12}O_{42}]^{10-}$ 结构

11.4.3　十九钨氧簇

将盐酸三乙醇胺和钨酸钠溶于水，用盐酸调 pH 值至 1.2，回流 3d，室温放置 2d 后可以得到灰绿色晶体[118]，其分子组成为 $(TEAH)_6[H_4W_{19}O_{62}]\cdot 6H_2O$，研究表明，十九钨氧簇存在两种异构体，均具有类 Dawson 型结构，其中一个钨氧八面体 WO_6 位于阴离子的中间，如图 11-31 所示。

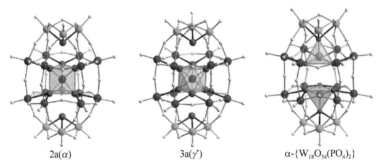

$2a(\alpha)$　　　　　$3a(\gamma^*)$　　　　$\alpha\text{-}\{W_{18}O_{54}(PO_4)_2\}$

图 11-31　$[H_4W_{19}O_{62}]^{6-}$ 结构

11.4.4　二十二钨氧簇

亚硫酸钠与沸腾的钨酸钠水溶液反应，通过调节溶液 pH 值和其他反应条件，最终可得到分子组成为 $Na_{12}[H_4W_{22}O_{74}]\cdot 31H_2O$ 的化合物[119]。晶体结构解析表明，阴离子具有 S 形结构，可看作由两个 $[H_4W_{11}O_{38}]^{6-}$ 缩合而成的，结构如图 11-32 所示。

11.4.5　二十四钨氧簇

钨酸铯在氩气氛下与钨酸水溶液在 20℃反应 2h，可得到分子组成为

$Cs_{24}W_{24}O_{84} \cdot 24H_2O$ 的化合物[120]，其结构如图 11-33 所示。

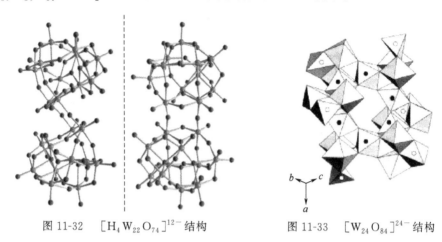

图 11-32　$[H_4W_{22}O_{74}]^{12-}$ 结构　　　　图 11-33　$[W_{24}O_{84}]^{24-}$ 结构

11.4.6　二十八钨氧簇

尽管二十八钨氧簇作为独立的阴离子尚未分离出来，但由其和稀土离子配位形成的 $[Ln_2(H_2O)_{10}W_{28}O_{93}(OH)_2]^{14-}$ 构成的化合物已经得到[121]。

钨酸钠溶于水，控制反应条件与稀土氧化物反应，可以得到由 $[Ln_2(H_2O)_{10}W_{28}O_{93}(OH)_2]^{14-}$ 构成的化合物。阴离子结构如图 11-34 所示，阴离子由两个 W_{11} 基团和一个 W_6 基团通过共顶点连接，再与两个稀土配位构成。

图 11-34　$[Ln_2(H_2O)_{10}W_{28}O_{93}(OH)_2]^{14-}$ 结构

11.4.7　三十四钨氧簇

亚硫酸钠搅拌下加入沸腾的钨酸钠溶液中，控制反应条件，最终得到分子组成为 $Na_{18}[H_{10}W_{34}O_{116}] \cdot 47H_2O$ 的化合物[119]。在该化合物中，三十四钨氧簇阴离子由两个 W_{11} 基团于反位通过桥氧与 $[H_2W_{12}O_{42}]^{10-}$ 基团相连，整个阴离子构成 S 形，如图 11-35 所示。

11.4.8 三十六钨氧簇

将三乙醇胺和钨酸钠溶于水中,用盐酸调 pH 值,加入连二亚硫酸钠,控制反应条件,最终可得到分子组成为 $(TEAH)_9Na_2\{(H_2O)_4K\subset[H_{12}W_{36}O_{120}]\}\cdot 17H_2O$ 的化合物[122]。在该化合物中三十六钨氧簇阴离子由三个 $[H_4W_{11}O_{38}]^{6-}$ 基团通过三个 W 原子相连,构成近似具有 C_{3V} 对称性的环状结构,钾离子位于环的中央,如图 11-36 所示。后续研究工作表明,环中央的钾离子可以被 Rb^+、Cs^+、NH_4^+、Sr^{2+}、Ba^{2+} 等离子取代[123]。

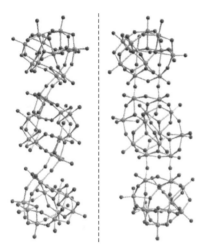
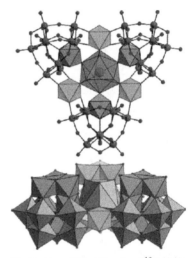

图 11-35 $[H_{10}W_{34}O_{116}]^{18-}$ 结构   图 11-36 $[H_{12}W_{36}O_{120}]^{12-}$ 结构

参考文献

[1] Day V W, Klemperer W G, Yaghi O M. Synthesis and characterization of a soluble oxide inclusion complex, $[CH_3CN\subset(V_{12}O_{32}^{4-})]$. J Am Chem Soc, 1989, 111: 5959-5961.

[2] Müller A, Rohlfing R, Krickemeyer E, Bögge H. Control of the linkage of inorganic fragments of V—O compounds: from cluster shells as carcerands via cluster aggregates to solid-state structures. Angew Chem Int Ed Engl, 1993, 32 (6): 909-911.

[3] Kawanami N, Ozeki T, Yagasaki A. NO^- anion trapped in a molecular oxide bowl. J Am Chem Soc, 2000, 122: 1239-1240.

[4] Hou D, Hagen K S, Hill C L. Tridecavanadate, $[V_{13}O_{34}]^{3-}$, a new high-potential isopolyvanadate. J Am Chem Soc, 1992, 114: 5864-5866.

[5] Pettersson L, Andersson I, Howarth O W. Tridecavanadate, $[H_{12}V_{13}O_{40}]^{3-}$. Inorg Chem, 1992, 31: 4032-4033.

[6] Hao N, Qin C, Xu Y, Wang E B, Li Y, Shen E, Xu L. A new pseudo Keggin-type

polyvanadate $(NH_4)_4[H_{12}V_{12}^{V}O_{36}(V^{IV}O_4)] \cdot 11H_2O$: hydrothermal synthesis and crystal structure. Inorg Chem Commun, 2005, 8: 592-595.

[7] Chen L, Jiang F, Lin Z, Zhou Y, Yue C, Hong M. A basket tetradecavanadate cluster with blue luminescence. J Am Chem Soc, 2005, 127: 8588-8589.

[8] Müller A, Krickemeyer E, Penk M, Walberg H Bögge. Spherical mixed-valence $[V_{15}O_{36}]^{5-}$, an example from an unusual cluster family. Angew Chem Int Ed Engl, 1987, 26 (10): 1045-1046.

[9] 邵美成, 冷晶, 曾华东, 潘左华, 唐有祺. 混合价聚十五钒酸盐 $[V_{15}O_{36}Cl](NH_4)_3Na_7 \cdot 30H_2O$ 的合成与晶体结构. 高等学校化学学报, 1990.11 (3): 280-285.

[10] Hayashi Y, Myakoshi N, Shinguchi T, Uehara A. A stepwise growth of polyoxovanadate by reductive coupling reaction with organometallic palladium complex: formation of $[\{(\eta^3-C_4H_7)Pd\}_2V_{12}O_{12}]^{2-}$, $[V_{10}O_{26}]^{4-}$ and $[V_{15}O_{36}(Cl)]^{4-}$. Chem Lett, 2001: 170-171.

[11] Yamase T, Ohtaka K. Photochemistry of polyoxovanadates. Part 1. Formation of the anion-encapsulated polyoxovanadate $[V_{15}O_{36}(CO_3)]^{7-}$ and electron-spin polarization of α-hydroxyalkyl radicals in the presence of alcohols. J Chem Soc Dalton Trans, 1994: 2599-2608.

[12] Chen L, Jiang F, Wu M, Li N, Xu W, Yan C, Yue C, Hong M. Half-open hollow cages of pentadecavanadate and hexadecavanadate compounds with large-O-V-O-V-windows. Cryst Growth Des, 2008, 8 (11): 4092-4099.

[13] Hou D, Hagen K S, Hill C L. Pentadecavanadate, $V_{15}O_{42}^{9-}$, a new highly condensed fully oxidized isopolyvanadate with kinetic stability in water. J Chem Soc Chem Commun, 1993: 426-428.

[14] Duraisamy T, Ojha N, Ramanan A, Vittal J J. New inorganic-organic hybird materials based on polyoxovanadates: $[Morp]_6[VO_4 \subset V_{14}O_{32}(OH)_6] \cdot 2H_2O$, $[Morp]VO_3$, $[HATM-CH_3]_4[H_2V_{10}O_{28}] \cdot 5H_2O$, and $[HMTA-H][HMTA-CH_2OH]_2[H_5V_{10}O_{28}]$ $6H_2O$. Chem Mater, 1999, 11: 2339-2349.

[15] Pavani K, Upreti S, Ramanan A. Two polyoxovanadate clusters templated through cysteamine. J Chem Sci, 2006, 118 (2): 159-164.

[16] Lin B, Liu S. First hexadecavanadate compound: hydrothermal synthesis and characterization of a three-dimensional framework $[\{Cu(1,2-pn)_2\}_7\{V_{16}O_{38}(H_2O)\}_2] \cdot 4H_2O$. Chem Commun, 2002: 2126-2127.

[17] Long D, Orr D, Seeber G, Kögerler P, Farrugia L J, Cronin L. The missing link in low nuclearity pure polyoxovanadate clusters: preliminary synthesis and structural analysis of a new $\{V_{16}\}$ cluster and related products. J Clust Sci, 2003, 14 (3): 313-324.

[18] Pan C, Xu J, Li G, Chu D, Wang T. A three-dimensinal framework of novel vanadium clusters bridged by $[Ni(en)_2]^{2+}$: $Ni(en)_3\{V_{11}^{IV}V_5^{V}O_{38}Cl[Ni(en)_2]_3\} \cdot 8.5H_2O$. Eur J Inorg Chem, 2003: 1514-1417.

[19] Cui X, Zheng S, Sun Y, Yang G. Hydrothermal synthesis and characterization of $[V_{16}O_{38}(Cl)][Cu(enMe)_2]_{3.5} \cdot 2H_2O$ containing $\{V_{16}O_{38}\}$ cluster. Chem Res Chinese U, 2004, 20 (3): 266-269.

[20] Dong B, Chen Y, Peng J, Kong Y, Han Z. Hydrothermal synthesis and crystal structure of a new mixed-valence mesostructured hexadecavanadate. J Mol Struct, 2005, 748: 171-176.

[21] Liu S, Xie L, Gao B, Zhang C, Sun C, Li D, Su Z. An organic-inorganic hybird material constucted from a three-dimensional coordination complex cationic framework and entrapped hexadecavanadate clusters. Chem Common, 2005: 5023-5025.

[22] Hayashi Y, Fukuyanma K, Takatera T, Uehara A. Synthesis and structure of a new reduced isopolyvanadate, $[V_{17}O_{42}]^{4-}$. Chem Lett, 2000: 770-771.

[23] Müller A, Penk M, Rohlfing R, Krickemeyer E, Döring J. Topologically interesting cages for negative ions with extremely high "coordination number": an unusual property of V—O cluster. Angew Chem Int Ed Engl, 1990, 29 (8): 926-927.

[24] Yamase T, Ohtaka K, Suzuki M. Structural characterization of spherical octadecavanadates encapsulating Cl^- and H_2O. J Chem Soc Dalton Trans, 1996, 3: 283-289.

[25] Müller A, Sessoli R, Krickemeyer E, Bögge H, Meyer J, Gattschi D, Pardi L, Westphal J, Hovemeier K, Rohlfing R, Döring J, Hellweg F, Beugholt C, Schmidtmann M. Polyoxovanadates: high-nuclearity spin clusters with interesting host-guest systems and different electron population. Synthesis, spin organization, magnetochemistry, and spectroscopic studies. Inorg Chem, 1997, 36: 5239-5250.

[26] Müller A, Döring J. Topologisch und elektronisch bemerkenswerte "reduzlerte" cluster des typs $[V_{18}O_{42}(X)]^{n-}$ ($X=SO_4$, VO_4) mit T_d-symmetrie und davon abgeleitete cluster $[V_{(18-p)}As_{2p}O_{42}(X)]^{m-}$ ($X=SO_3$, SO_4, H_2O; $p=3$, 4). Z Anorg Allg Chemie, 1991, 595: 251-274.

[27] Khan M I, Yohannes E, Doedens R J. A novel series of materials composed of arrays of vanadium oxide container molecules, $\{V_{18}O_{42}(X)\}$ ($X=H_2O, Cl^-, Br^-$): synthesis and characterization of $[M_2(H_2N(CH_2)_2NH_2)_5][\{M_2(H_2N(CH_2)_2NH_2)_2\}_2V_{18}O_{42}(X)] \cdot 9H_2O$ ($M=Zn$, Cd). Inorg Chem, 2003, 42 (9): 3125-3129.

[28] 崔小兵, 郑寿添, 丁兰, 丁红, 杨国昱. $[Cu(enMe)_2]_4[H_5V_{18}^{IV}O_{42}(Cl)] \cdot 8H_2O$: 由金属-氧簇和金属配合物构筑的微孔材料的合成和结构. Chin J Struct Chem, 2003, 22 (4): 491-494.

[29] Müller A, Krickemeyer E, Penk M, Rohlfing R, Armatage A, Bögge H. Template-controlled formation of cluster shells or a type of molecular recognition: synthesis of $[HV_{22}O_{54}(ClO_4)]^{6-}$ and $[H_2V_{18}O_{44}(N_3)]^{5-}$. Angew Chem Int Ed Engl, 1991, 30 (12): 1674-1677.

[30] Yamase T, Suzuki M, Ohtaka K. Structures of photochemically prepared mixed-valence polyoxovanadate clusters: oblong $[V_{18}O_{44}(N_3)]^{14-}$, super Keggin $[V_{18}O_{42}(PO_4)]^{11-}$

and doughunt-shaped $[V_{12}B_{32}O_{84}Na_4]^{15-}$ anions. J Chem Soc Dalton Trans, 1997: 2463-2472.

[31] Müller A, Penk M, Krickemeyer E, Bögge H, Walberg H. $[V_{19}O_{41}(OH)_9]^{8-}$, an ellipsoid-shaped cluster anion belonging to the unusual family of V^{IV}/V^V oxygen clusters. Angew Chem Int Ed Engl, 1998, 27 (12): 1719-1921.

[32] Suber L, Bonamico M, Fares V. Synthesis, magnetism, and X-ray molecular structure of the mixed-valence vanadium (IV/V)-oxygen cluster $[VO_4 \subset (V_{18}O_{45})]^{9-}$. Inorg Chem, 1997, 36: 2030-2033.

[33] Ishaque M K, Yohannes E, Powell D. Vanadium oxide clusters as building blocks for the synthasis of metal oxide surfaces and franmework materials: synthesis and X-ray crystal structure of $[H_6Mn_3V_{15}^{IV}V_4^VO_{46}(H_2O)_{12}] \cdot 3H_2O$. Inrog Chem, 1999, 38: 212-213.

[34] Müller A, Rohlfing R, Döring J, Penk M. Formation of a cluster sheath around a central cluster by a "self-organization process": the mixed valence polyxovanadate $[V_{34}O_{82}]^{10-}$. Angew Chem Int Ed Engl, 1991, 30 (5): 588-590.

[35] Maekawa M, Ozawa Y, Yagasaki A. Icosaniobate: a new member of the isoniobate family. Inorg Chem, 2006, 45: 9608-9609.

[36] Bontchev R P, Nyman Ma. Evolution of polyoxoniobate cluster anions. Angew Chem Int Ed, 2006, 45: 6670-6672.

[37] Niu J, Ma P, Niu H, Li J, Zhao J, Song Y, Wang J. Giant polyniobate clusters based on $[Nb_7O_{22}]^{9-}$ units derived from a Nb_6O_{19} precursor. Chem Eur J, 2007, 13: 8739-8748.

[38] Wang J, Niu H, Niu J. Preparation, crystal structure, and characterization of an inorganic-organic hybird polyoxoniobate $[Cu(en)_2]_3[Cu(en)_2(H_2O)]_{1.5}[K_{0.5}Nb_{24}O_{72}H_{14.5}] \cdot 2 \cdot 25H_2O$. J Chem Sci, 2008, 120 (3): 309-313.

[39] Tsunashima R, Long D, Miras H N, Gabb D, Pradeep C P, Cronin L. The construction of high-nuclearity isopolyoxoniobates with pentagonal building blocks: $[HNb_{27}O_{76}]^{16-}$ and $[H_{10}Nb_{31}O_{93}(CO_3)]^{23-}$. Angew Chem Int Ed, 2009, 48: 113-116.

[40] Chae H K, Klemperer W G, Páze Loyo D E, Day V W, Eberspacher T A. Synthesis and structure of a high-nuclearity oxomolybdenum (V) complex, $[(C_5Me_5Rh^{III})_8(Mo_{12}^VO_{36})(Mo^{VI}O_4)]^{2+}$. Inorg Chem, 1992, 31: 3187-3189.

[41] Jin X, Tang K, Ni H, Tang Y. Synthesis and crystal structure of a novel keggin-type isopolyoxomolybdate $[Bu_4N]_6[H_3O]_2[Mo_{13}O_{40}]_2$. Polyhedron, 1994, 15-16: 2439-2441.

[42] Khan M I, Müller A, Dillinger S, Bögge H, Chen Q, Zubieta J. Cation inclusion within the mixed-valence polyanion cluster $[(Mo^{VI}O_3)_4Mo_{12}^VO_{28}(OH)_{12}]^{8-}$: syntheses and structures of $(NH_4)_7[NaMo_{16}(OH)_{12}O_{40}] \cdot 4H_2O$ and $(Me_2NH_2)_6[H_2Mo_{16}(OH)_{12}O_{40}]$. Angew Chem Int Ed Engl, 1993, 32 (12): 1780-1782.

[43] Khan M I, Chen Q, Salta J, O'Connor C J, Zubieta J. Retention of structural cores in the synthesis of high-nuclearity polyoxoalkoxomolybdate clusters encapsulating [Na

$(H_2O)_3]^+$ and $[MoO_3]$ moieties. Hydrothermal syntheses and structures of $(NH_4)_7$ $[NaH_{12}Mo_{16}O_{52}] \cdot 4H_2O$ and $(Me_3NH)_4K_2[H_{14}Mo_{16}O_{52}] \cdot 8H_2O$ and their structural relationships to the class of superclusters $[XH_nMo_{42}O_{109}\{(OCH_2)_3CR\}_7]^{m-}$ (X=Na $(H_2O)^{3+}$：$n=13$，$m=9$；$n=15$，$m=7$. X=MoO_3：$n=14$，$m=9$；$n=13$，$m=10$). Inorg Chem，1996，35：1880-1901.

[44] Long D，Kögerler P，Farrugia L J，Cronin L. Restraining symmerty in the formation of small polyoxomolybdates: building blocks of unprecedented topology resulting from "shrink-wrapping" $[H_2Mo_{16}O_{52}]^{10-}$-type clusters. Angew Chem Int Ed，2003，42：4180-4183.

[45] Brown I D. in Structure and Bonding in Crystals, Vol. II (Eds.: M. O' Keeffe, A. Navrotsky), New York: Academic Press, 1981: 1-30.

[46] Tytko K，Schönfeld B，Buss B，Glemser O. A macroisopolyanion of molybdenum: $Mo_{36}O_{112}^{8-}$，Angew Chem Internat Edit，1973，12 (4)：330-332.

[47] Paulat-Böschen I. X-ray crystallographic determination of the structure of the isopolyanion $[Mo_{36}O_{112}(H_2O)_{16}]^{8-}$ in the compound $K_8[Mo_{36}O_{112}(H_2O)_{16}] \cdot 36H_2O$. J C S Chem Comm，1979：780-782.

[48] Krebs B，Paulat-Böschen I. The structure of the potassium isopolymolybdate $K_8[Mo_{36}O_{112}(H_2O)_{16}] \cdot nH_2O(n=36, \cdots, 40)$. Acta Cryst, 1982, B38: 1710-1718.

[49] Zhang S W，Liao D Q，Shao M C，Tang Y Q. X-ray crystal structure of the unusual clathrate compound，$[Mo_{36}O_{110}(NO)_4(H_2O)_{14}] \cdot 52H_2O$. J of the Chem Soc，Chem Comm，1986：835-836.

[50] Liu Guang，Wei Yong Ge，Yu Qing，Liu Qun，Zhang Shi Wei. Polyoxmetalate chain-like polymer: the synthesis and crystal structure of a 1D compound，$[Mo_{36}O_{108}(NO)_4(MoO)_2La_2(H_2O)_{28}]_n \cdot 56nH_2O$. Inorg Chem Commun，1999，2：434-437.

[51] Yang W，Lu C，Zhan X，Zhang Q，Liu J，Yu Y. A new member of giant polyoxomolybdate containing bridging hydroxylamine moieties: synthesis and crystal structure of $(NH_4)_4Na_2[Mo_{36}^{VI}O_{108}(NH_2OH)_2(OH)_6(H_2O)_{12}] \cdot 35H_2O$. J Clust Sci，2003，14 (3)：391-403.

[52] Izarova N V，Sokolov M N，Dolgushin F M，Antipin M Y，Fenske D，Fedin V P. A novel two-dimensional framework solid composed of nanosized molybdenum-oxide molecules: synthesis and characterization of $[\{Gd(H_2O)_5\}_4\{Mo_{36}(NO)_4O_{108}(H_2O)_{16}\}] \cdot 34H_2O$. C R Chim，2005，8：1922-1926.

[53] Izarova N V，Sokolov M N. Samsonenko D G，Rothenberger A，Naumov D Y，Fenske D，Fedin V. One-，two-，and three-dimensional coordination polymers built from large Mo_{36}-polyoxometalate anionic units and lanthanide cations. Eur J Inorg Chem，2005：4985-4996.

[54] Yamase T，Ishikawa E. Photochemical self-assembly reaction of β-$[Mo_8O_{26}]^{4-}$ to mixed-valence cluster $[Mo_{37}O_{112}]^{26-}$ in aqueous media. Langmuir，2000，16：9023-9030.

[55] Khan M I, Zubieta J. A high-nuclearity polyoxoalkoxomolybdate cluster encapsulating a $[Na(H_2O)_3]^+$ moiety. Hydrothermal synthesis and structure of $[Na(H_2O)_3 H_{15} Mo_{42} O_{109} \{(OCH_2)_3 CCH_2 OH\}_7]^{7-}$. J Am Chem Soc, 1992, 114: 10058-10059.

[56] Yang W, Lu C, Lin X, Zhuang H. Novel acetate polyoxomolybdate "host" accommodating a zigzag-chainlike "guest" of five edge-shared sodium cations: $Na_{21} \{[Na_5 (H_2O)_{14}] \subset [Mo_{46} O_{134} (OH)_{10} (\mu\text{-}CH_3 COO)_4]\} CH_3 COONa \cdot \sim 95 H_2 O$. Inorg Chem, 2002, 41: 452-454.

[57] Yang W, Lu C, Lin X, Zhuang H. A novel acetated 54-member crown-shaped polyoxomolybdate with unprecedented structural features: $Na_{26} [\{Na(H_2O)_2\}_6 \{(\mu_3\text{-}OH)_4 Mo_{20}^V Mo_{34}^{VI} O_{164} (\mu_2\text{-}CH_3 COO)_4\}] \cdot \sim 120 H_2 O$. Chem Commun, 2000: 1623-1624.

[58] Cronin L, Beugholt C, Krickemeyer E, Schmidtmann M, Bögge H, Kögerler P, Kim T, Luong K, Müller A. "Molecular symmetry breakers" generating metal-oxide-based nanoobject fragments as synthons for complex structures: $[\{Mo_{128} Eu_4 O_{388} H_{10} (H_2O)_{81}\}_2]^{20-}$, a giant-cluster dimer. Angew Chem Int Ed, 2002, 41 (15): 2805-2808.

[59] Müller A, Krickemeyer E, Bögge H, Schmidmann M, Peters F. Organizational forms of matter: an inorganic super fullerence and keplerate based on molybdenum oxide. Angew Chem Int Ed, 1998, 37 (24): 3359-3363.

[60] Müller A, Beugholt C, Koop M, Das S K, Schmidtmann M, Bögge H. Facile and optimized syntheses and structures of crystalline molybdenum blue compounds including one with an interesting high degree of defects: $Na_{26} [Mo_{142} O_{432} (H_2O)_{58} H_{14}] \cdot ca. 300 H_2 O$ and $Na_{16} [(MoO_3)_{176} (H_2O)_{63} (CH_3 OH)_{17} H_{16}] \cdot ca. 600 H_2 O \cdot ca. 6 CH_3 OH$. Z Anorg Allg Chem, 1999, 625: 1960-1962.

[61] Yamase T, Prokop P V. Photochemical formation of tire-shaped molybdenum blues: topology of a defect anion, $[Mo_{142} O_{432} H_{28} (H_2O)_{58}]^{12-}$. Angew Chem Int Ed, 2002, 41 (3): 466-469.

[62] Müller A, Krickemeyer E, Bögge H, Schmidmann M, Peters F, Menke C, Meyer J. An unusual polyoxomolybdate: giant wheels linked to chains. Angew Chem Int Ed Engl, 1997, 36 (5): 484-486.

[63] Müller A, Das S K, Fedin V P, Krickemeyer E, Beugholt C, Bögge H, Schmidmann M, Hauptfleisch B. Rapid and simple isolation of the crystalline molybdenum-blue compounds with discrete and linked nanosized ring-shaped anions: $Na_{15} [Mo_{126}^{VI} Mo_{28}^V O_{462} H_{14} (H_2O)_{70}]_{0.5} [Mo_{124}^{VI} Mo_{28}^V O_{457} H_{14} (H_2O)_{68}]_{0.5} \cdot ca. 250 H_2 O$. Z Anorg Allg Chem, 1999, 625: 1187-1192.

[64] Müller A, Krickemeyer E, Bögge H, Schmidtmann M, Beugholt C, Das S K, Peters F. Giant ring-shaped building blocks linked to form a layered cluster network with nanosized channels: $[Mo_{124}^{VI} Mo_{28}^V O_{429} (\mu_3\text{-}O)_{28} H_{14} (H_2O)_{66.5}]^{16-}$. Chem Eur J, 1999, 5: 1496-1502.

[65] Müller A, Krickemeyer E, Meyer J, Bögge H, Peters F, Plass W, Diemann E,

Dillinger S, Nonnenbruch F, Randerath M, Menke C. [$Mo_{154}(NO)_{14}O_{420}(OH)_{28}(H_2O)_{70}$]$^{(25\pm5)-}$: a water-soluble big wheel with more than 700 atoms and a relative molecular mass of about 24000. Angew Chem Int Ed Engl, 1995, 34: 2122-2124.

[66] Müller A, Meyer J, Krickemeyer E, Diemann E. Molybdenum blue: a 200 year old mystery unveiled. Angew Chem Int Ed Engl, 1996, 35 (11): 1206-1208.

[67] Shishido S, Ozeki T. The pH dependent nuclearity variation of {Mo_{154-x}}-type polyoxomolybdates and tectonic effect on their aggregations. J Am Chem Soc, 2008, 130: 10588-10595.

[68] Müller A, Krickemeyer E, Bögge H, Schmidtmann M, Beugholt C, Kögerler P, Lu C. Formation of a ring-shaped reduced "metal oxide" with the simple composition [$(MoO_3)_{176}(H_2O)_{80}H_{32}$]. Angew Chem Int Ed, 1998, 37: 1220-1223.

[69] Jiang C, Wei Y, Liu Q, Zhang S, Shao M, Tang Y. Self-assembly of a novel nanoscale fiant cluster: [$Mo_{176}O_{496}(OH)_{32}(H_2O)_{80}$]. Chem Commun, 1998: 1937-1938.

[70] Imai H, Akutagawa T, Kudo F, Ito M, Toyoda K, Noro S, Cronin L, Nakamura T. Structure, magnetism, and ionic conductivity of the gigantic {Mo_{176}}-wheel assembly: $Na_{15}Fe_3Co_{16}$[$Mo_{176}O_{528}H_3(H_2O)_{80}$]$Cl_{27}$ · $450H_2O$. J Am Chem Soc, 2009, 131: 13578-13579.

[71] Müller A, Koop M, Bögge H, Schmidtmann M, Beugholt C. Exchanged ligands on the surface of a giant cluster: [$(MoO_3)_{176}(H_2O)_{63}(CH_3OH)_{17}H_n$]$^{(32-n)-}$. Chem Commun, 1998: 1501-1502.

[72] Müller A, Shah S Q N, Bögge H, Schmidtmann M. Molecular growth from a Mo176 to a Mo248 cluster. Nature, 1999, 397: 48-50.

[73] Müller A, Beckmann E, Bögge H, Schmidtmann M, Dress A. Inorganic chemistry goes protein size: a Mo368 nano-hedgehog initiating nanochemistry by symmetry breaking. Angew Chem Int Ed, 2002, 41 (7): 1162-1167.

[74] Ortiz-Acosta D, Feller R K, Scott B L, Del Sesto R E. Isolation of an asymmetric lanthanide polyoxometalate, $Na_{12}H$[$(W_5O_{18})Tb(H_2W_{11}O_{39})$] · $42H_2O$, containing two distinct isopalyanions. J Chem Crystallogr, 2012, 42: 654-655.

[75] Freedman M L. The tungstic acids. J Am Chem Soc, 1959, 81 (15): 3834-3839.

[76] Signer R, Gross H. Über den bau einiger heteropolysäuren. Helv Chim Acta, 1934, 17: 1076-1080.

[77] Pope M T V, Arga G M Jr. Proton magnetic resonance of aqueous metatungstate ion: evidence for two central hydrogen atoms. Chem Commun, 1966: 653-654.

[78] Launay J, Boyer M, Chauveau F. High resolution PMR of several isopolytungstates and related compounds. J Inorg Nucl Chem, 1976, 38: 243-247.

[79] Fuchs J, Flindt E. Preparation and structure investigation of polytungstates. Z Natuiforsch, 1979, 34b: 412-422.

[80] Asami M, Ichida H, Sasaki Y. The structure of hexakis (tetramethylammonium) di-

hydrogendodecatungstate enneahydrate, $[(CH_3)_4N]_6[H_2W_{12}O_{40}] \cdot 9H_2O$. Acta Cryst, 1984, C40: 35-37.

[81] Zavalij P, Guo J, Whittingham M S, Jacobson R A, Pecharsky V, Bucher C K, Hwu S. Keggin cluster formation by hydrothermal reaction of tungsten trioxide with methyl substituted ammonium: the crystal structure of two novel compounds, $[NH_2(CH_3)_2]_6H_2W_{12}O_{40} \cdot 4H_2O$ and $[NH_2(CH_3)_4]_6H_2W_{12}O_{40} \cdot 2H_2O$. J Solid State Chem, 1996, 123: 83-92.

[82] Launay J P. Reduction de L'ion metatungstate: stades eleves de reduction de $H_2W_{12}O_{40}^{6-}$, derives de de L'ion $HW_{12}O_{40}^{7-}$, et discussion generale. J Inorg Nucl Chem, 1976, 38: 807-816.

[83] Jeannin Y, Launay J P, Seid Sedjadi M A. Crystal and molecular structure of the six-electron-reduced form of metatungstate $Rb_4H_8[H_2W_{12}O_{40}] \cdot 18H_2O$: occurrence of metal-metal bonded subcluster in a heteropalyanion framework. Inorg Chem, 1980, 19: 2933-2935.

[84] Colette Boskovic, Maruse Sadek, Robert T C Brownlee, Alan M Bond, Anthony G Wedd. Electrosynthesis and solution structure of six-electron reduced forms of metatungstate, $[H_2W_{12}O_{40}]^{6-}$. J Chem Soc, Dalton Trans, 2001: 187-196.

[85] Yao W, Wang X, Xie H, Ma R, Yu Y, Li Y. New supramolecular networks based on paratungstate with N-donor bridging ligands. Chem Res Chinese Universities, 2012, 28 (30): 382-386.

[86] Niu J, Zhao J, Wang J, Bo Y. Syntheses, spectroscopic characterization, thermal behavior, electrochemistry and crystal structures of two novel pyridine metatungstates. J Coord Chem, 2004, 57 (11): 935-946.

[87] Niu J, Wang Z, Wang J. Hydrothermal synthesis and crystal structure of a nonel metatungstate $[Cu(2,2'-bipy)_3]_2H_2[H_2W_{12}O_{40}] \cdot 4.5H_2O$. J Coord Chem, 2004, 57 (5): 411-416.

[88] Zhou D, Ren Q, Zhao J, Wang J. Synthesis, characterization and crystal structure of an inorganic-organic composite metatungstate $[Co(4,4'-bipy)_2(H_2O)_4](4,4'-H_2bipy)_2[H_2W_{12}O_{40}] \cdot 5.5H_2O$. Chinese J Struct Chem, 2007, 26 (11): 1287-1292.

[89] Yan B, Li Y, Zhao H, Pan W, Parkin S. Polyoxometalates functioned as ligands: synthesis and crystal structure of a new hybrid compound constructed from metatungstate and metal complex units. Inorg Chem Commun, 2009, 12: 1139-1141.

[90] Yang X, Chen Y, Zhang C, Kong Q, Yao F. Nine-coordinated Ag^+ ion in a diamondoid 3D all-inorganic framework constructed from α-metatungstate anions and Ag^+ ions: synthesis and structure of $[\{Ag(H_2O)_3(Bu^iNH_2)_4\}_2Na_2H_2(H_2W_{12}O_{40})] \cdot 4H_2O$. Solid State Sci, 2011, 13: 476-479.

[91] Wang J, Ren Q, Zhao J, Niu J. An organic-inorganic hybrid polyoxometalate based on the dodecatungstate building block: $[Ni(phen)(H_2O)_3]_2[Ni(H_2O)_5][H_2W_{12}W_{40}] \cdot$

[92] Zhang C, Pang H, Wang D, Chen Y. Two α-metatungstate compounds containing supramolecular helical chains. J Coord Chem, 2012, 63 (4): 568-578.

[93] Sun P, Liu S, Feng D, Ma F, Zhang W, Ren Y, Cao J. A novel organic-inorganic hybrid based on a dinuclear copper (II)-oxalate complex, a α-metatungstate cluster $[H_2W_{12}O_{40}]^{6-}$ with catalytic activity in H_2O_2 decomposition. J Mol Struct, 2010, 968: 89-92.

[94] Streb C, Ritchie C, Long D, Kögerler P, Cronin L. Modular assembly of a functional polyoxometalate-based open framework constructed from unsupported $Ag^I \cdots Ag^I$ interactions. Angew Chem Int Ed, 2007, 46: 7579-7582.

[95] Nandini Devi R, Burkholder E, Zubieta J. Hydrothermal synthesis of polyoxometalate clusters, surface-modified with M(II)-organonitrogen subunits. Inorg Chim Acta, 2003, 348: 150-156.

[96] Zhang P, Shen X, Han Z, Tian A, Pang H, Sha J, Chen Y, Zhu M. A twofold interpenetrating framework based on the α-metatungstates. J Solid State Chem, 2009, 182: 3399-3405.

[97] Hu M, Chen Y, Zhang C, Kong Qi. High-dimensional frameworks dependent on coordination mode of ligang controlled by acidity of reaction. Syntheses, structures, magnetic and fluorescence properties of eight new compounds. Cryst Eng Comm, 2010, 12, 1454-1460.

[98] Chen H, Ding Y, Wang E, Xu X, Qi Y, Chang S, Li Y. Strutural characterization of two lanthanide complexes attached to $[H_2W_{12}O_{40}]^{6-}$. Trans Met Chem, 2008, 33: 341-346.

[99] Pang H, Chen Y, Meng F, Shi D. Assembly of three novel 2D frameworks with helical chains based on $[H_2W_{12}O_{40}]^{6-}$ clusters and lanthanide-organic complexes. Inorg Chim Acta, 2008, 361: 2508-2514.

[100] Pang H, Chen Y, Meng F, Shi D. Two novel one-dimensional inorganic-organic-hybrids constructed from $[H_2W_{12}O_{40}]^{6-}$ clusters and lanthanide-organic complexes: $[(C_5H_5N-CO_2)_2Ln(H_2O)_3]_2[H_2W_{12}O_{40}] \cdot nH_2O$. Z Naturforsch, 2008, 63b: 16-22.

[101] Liu D, Chen Y, Zhang C, Meng H, Zhang Z, Zhang C. Effect of ring coordination of pyridine-3,5-dicarboxylate and metatungstate to Ln ions on metatungstate strutures: synthesis, structure and optical property of four new compounds. J Solid State Chem, 2011, 184: 1355-1360.

[102] Zhang Z. Self-assembly of isopolyanion clusters and lanthanide-organic units into 2D layers: $(NH_4)_2\{[Ln_2(C_7H_4NO_4)_2(H_2O)][(H_2W_{12}O_{40})]\} \cdot nH_2O$ (Ln=Gd, Tb, Ho). J Clust Sci, 2011, 22: 705-714.

[103] Meng H, Chen Y, Liu M, Liu D, Zhang Z, Zhang C. Liu S. Syntheses, structures

and properties of three-dimensional organic-inorganic hybrids based on α-metatungstate. Inorg Chim Acta, 2012, 387: 8-14.

[104] Allmann R. Die Struktur des ammoniumparawolframates $(NH_4)_{10}[H_2W_{12}O_{42}] \cdot 10H_2O$. Acta Crystallogr Sect B, 1971, B27: 1393-1404.

[105] D'Amour H, Allmann R. Crystal structure of ammonium paratungstate tetrahydrate $(NH_4)_{10}(H_2W_{12}O_{42}) \cdot 4H_2O$. Z Kristallogr, 1972, 136: 23-47.

[106] D'Amour H, Allmann R. Crystal structure of disodium octammonium paratungstate dodecahydrate, $Na_2(NH_4)_8[H_2W_{12}O_{42}] \cdot 12H_2O$. Z Kristallogr, 1973, 138: 5-18.

[107] Aveibach-Pouchot M T, Tordjman I, Durif A, Guitel J C. Structure d'un paratungstate d'ammonium$(NH_4)_6H_6W_{12}O_{42} \cdot 10H_2O$. Acta Crystallogr Sect B, 1979, B35: 1675-1677.

[108] Evans H T Jr, Rollins O W. Sodium paradodecatungstate 20-hydrate. Acta Crystallogr Sect B, 1976, B32: 1565-1567.

[109] Evans H T Jr, Prince E. Location of internal hydrogen atoms in the paradodecatungstate polyanion by neutron diffraction. J Am Chem Soc, 1983, 105: 4838-4839.

[110] Cruywagen J J, Van Der Merwe I F J, Nassimbeni L R, Niven M L, Symonds E A. Crystal and molecular structure of sodium paratungstate 26 hydrate. J Crystal Spect Res, 1986, 16 (4): 525-535.

[111] Han W, Hibino M, Kudo T. Synthesis of the hexagonal form of tungsten trioxide from peroxopolytungstate via ammonium paratungstate dscahydrate. Bull Chem Soc Jpn, 1998, 71: 933-937.

[112] Nolan A L, Allen C C, Burns R C, Lawrance G A, Wilkes E N, Hambley T W. Crystal structure of $Na_9[H_3W_{12}O_{42}] \cdot 24H_2O$, a compound containing the protonataed paratungstate B anion ("Acid Paratungstate"), and cyclic voltammetry of acidified $[H_2W_{12}O_{42}]^{10-}$ solutions. Aust J Chem, 1999, 52: 955-963.

[113] Yuan L, Qin C, Wang X, Li Y, Wang E. Transition of classic decatungstate to paratungstate: synthesis structure and luminescence properties of two paratungstate-based 3-D compounds. Z Naturforsch, 2008, 63b: 1175-1180.

[114] He L, Lin B, Liu X, Huang X, Feng Y. A novel layer formed by paradodecatungstate clusters and $\{Cu(en)_2\}^{2+}$ bridging groups: synthesis and characterization of $[\{Cu(en)_2\}_4(H_4W_{12}O_{42})] \cdot 9H_2O$. Solid State Sci, 2008, 10: 237-243.

[115] Yuan L, Qin C, Wang X, Wang E, Li Y. A new series of polyoxometalate compounds built up of paradodecatungstate anions and transition metal/alkaline-earth metal cations. Solid State Sci, 2008, 10: 967-975.

[116] Radio S V, Kryuchkov M A, Zavialova E G, Baumei V N, Shishkin O V, Rozantsev G M. Equilibrium in the acidified aqueous solutions of tungstate anion: synthesis of Co (Ⅱ) paratungstate B $Co_5[W_{12}O_{40}(OH)_2] \cdot 37H_2O$. J Coord Chem, 2010, 63 (10): 1678-1689.

[117] Radio S V, Rozantsev G M, Baumei V N, Shishkin O V. Crystal structure of nickel paratungstate B $Ni_5[W_{12}O_{40}(OH)_2] \cdot 37H_2O$. J Struct Chem, 2011, 52 (1): 111-117.

[118] Long D, Kögerler P, Parenty A D C, Fielden J, Cronin L. Discovery of a family of isopolyoxotungstates $[H_4W_{19}O_{62}]^{6-}$ encapsulating a $\{WO_6\}$ moiety within a $\{W_{18}\}$ dawson-like cluster cage. Angew Chem Int Ed, 2006, 45: 4798-4803.

[119] Miras H N, Yan J, Long D, Cronin L. Structural evolution of "S"-shaped $[H_4W_{22}O_{74}]^{12-}$ and "§"-shaped $[H_{10}W_{34}O_{116}]^{18-}$ isopolyoxotungstate clusters. Angew Chem Int Ed, 2008, 47: 8420-8423.

[120] Brudgam I, Funchs J, Hartl H, Palm R. Two new isopolyoxotungstates (Ⅵ) with the empirical composition $Cs_2W_2O_7 \cdot 2H_2O$ and $Na_2W_2O_7 \cdot H_2O$: an icosatetratungstate and a polymeric compound. Angew Chem Int Ed, 1998, 37 (19): 2668-2671.

[121] Ismail A H, Bassil B S, Suchopar A, Kortz U. Synthesis and structural characterization of the 28-isopolyoxotungstates fragment $[H_2W_{28}O_{93}]^{20-}$ stabilized by two external lanthanide ions $[Ln_2(H_2O)_{10}W_{28}O_{93}(OH)_2]^{14-}$. Eur J Inorg Chem, 2009: 5247-5252.

[122] Long D, Abbas H, Kögerler P, Cronin L. A high-nuclearity "celtic-ring" isopolyoxotungstates, $[H_{12}W_{36}O_{120}]^{12-}$, that captures tarce potassium ions. J Am Chem Soc, 2004, 126: 13880-13881.

[123] Long D, Brücher O, Streb C, Cronin L. Inorganic crown: the host-guest chemistry of a high nuclearity "celtic-ring" isopolyoxotungstate $[H_{12}W_{36}O_{120}]^{12-}$. Dalton Trans, 2006: 2852-2860.

索 引

B

八钼氧簇 139
八钼氧簇 α 异构体 107
八钼氧簇 β 异构体 111
八钨氧簇 140

C

重氮烷衍生物 74，80

D

氮烯衍生物 72

E

二钒氧簇阴离子 4
二聚八面体 4
二聚八面体结构 8
二聚钒氧簇离子 4
二十八钨氧簇 208
二十二钒氧簇 194
二十二钨氧簇 207
二十铌氧簇阴离子 195
二十七铌氧簇 196
二十四铌氧簇阴离子 196
二十四钨氧簇 207

J

角型结构 15
金属有机衍生物 75，81
聚四钼氧簇阴离子 25
聚四钨氧簇阴离子 31

K

开环结构 15

L

镧系 45
六钒氧簇核 52
六钼氧簇 77
六钼氧簇阴离子 71
六铌氧簇 77
六铌氧簇钾盐 59
六铌氧簇钠盐 59
六铌氧簇阴离子 61，66
六钽氧簇 77
六钽氧簇阴离子 64，66
六钨氧簇 77

Q

七钼氧簇 95
七铌氧簇 94
七钨氧簇阴离子 100

S

三角双锥构型 40
三角双锥结构 38
三钼氧簇 16
三十六钼氧簇钾盐 200
三十六钨氧簇 209
三十七钼氧簇阴离子 201
三十四钒氧簇阴离子 195
三十四钨氧簇 208
三十一铌氧簇 196
三钨氧簇 16
十八钒氧簇阴离子 192
十二钒氧簇阴离子 183
十二钨氧簇 205

十钒氧簇阴离子 151
十九钒氧簇阴离子 193
十九钨氧簇 207
十六钒氧簇阴离子 189
十铌氧簇阴离子 165
十七钒氧簇阴离子 192
十三钒氧簇阴离子 185
十四钒氧簇阴离子 186
十钨氧簇四丁基铵盐 170
十五钒氧簇阴离子 186
十一钨氧簇阴离子 205
双八面体结构 7
四钒氧簇 21
四十二钼氧簇阴离子 201
四十六钼氧簇 201
四钨氧簇阴离子 28

W

无色晶体 30
五钒氧簇阴离子 37
五十四钼氧簇 201
五钨酸根 43
五钨氧簇阴离子 44

X

线型结构 16

Y

亚氨基衍生物 74
亚胺衍生物 79
亚肼衍生物 73
亚硝基衍生物 75,80
一百二十八钼氧簇阴离子 202
一百七十六钼氧簇阴离子 205
一百四十二钼氧簇 203

其它

η 构型 138
θ 构型 138
α 异构体 106
β 异构体 111
γ 异构体 125
δ 异构体 133
ε 异构体 135
ζ 异构体 137